616.99

D1281904

Cancer Happens

COMING OF AGE WITH CANCER

R E B E C C A   G I F F O R D

A CAPITAL DISCOVERIES BOOK

CAPITAL
BOOKS, INC.
Sterling, Virginia

Capital Books, Inc.
P.O. Box 605
Herndon, Virginia 20172-0605

ISBN 1-931868-13-1 (alk.paper)

Library of Congress Cataloging-in-Publication Data

Gifford, Rebecca
    Cancer happens : coming of age with cancer / Rebecca
    Gifford.—1st ed.
        p. cm. — (A Capital discoveries book)
    ISBN 1–931868–13–1
        1. Gifford, Rebecca—Health. 2. Lymphomas—Patients—
United States—Biography. I. Title. II. Series.

RC280.L9 G544 2003
362.1'9699446—dc21
                                                    2002190773

Printed in the United States of America on acid-free paper that meets the American National Standards Institute Z39-48 Standard.

First Edition

10 9 8 7 6 5 4 3 2 1

for my family, who took care of me

contents

foreword ix

acknowledgments xiii

introduction still life #1 1

chapter one cheating death 5

chapter two normal 13

chapter three a steroid buzz 23

chapter four hair stories 35

chapter five we've all been there 45

chapter six full support without pantyhose 53

chapter seven praying on my conscience 63

chapter eight the rebellion 71

chapter nine french kissing 79

chapter ten   we are how we live   87

chapter eleven   first steps   97

chapter twelve   next steps   107

chapter thirteen   homesick   113

chapter fourteen   is dad always right?   123

chapter fifteen   house of prayer   131

chapter sixteen   what was the question?   141

chapter seventeen   hi, my name is rebecca . . .   149

chapter eighteen   sitting beside me   157

chapter nineteen   back to normal, already?   167

chapter twenty   casualties of war   175

chapter twenty-one   goin' to graceland   183

chapter twenty-two   survival guide   191

chapter twenty-three   something to look forward to   207

chapter twenty-four   not quite peace yet, but getting there   219

appendix   cancer resource list   229

foreword

I met Rebecca Gifford on July 27, 1997, in a patient examination room at Memorial Sloan-Kettering Cancer Center in New York City. She was a twenty-five-year-old woman who had just moved from Cincinnati, Ohio, and was transferring her medical care to my supervision in New York. She impressed me at the time as being tremendously bold and independent; few cancer patients, especially a single twenty-five-year-old, would leave their immediate family and friends and embark on a "new life" so soon after completing aggressive treatment for non-Hodgkin's lymphoma.

I assumed care of Rebecca and had the opportunity to get to know her over the next few years—at least I thought I knew her. In early 1999 I helped start oncology.com, an online resource and information service for cancer patients, their families and friends, and healthcare professionals. It grew to become one of the largest online healthcare communities (while Internet-related companies were still thriving). I had recently read some of Rebecca's writings and decided to submit them to the oncology.com editor in chief. I sent him an e-mail together with the writings to ask if he thought that writings like this would have general interest. His response came later that day: "YES, no question about it; this is great stuff!"

He suggested Rebecca write a personal experience column for the site. Rebecca agreed and began to chronicle her past experiences and her personal fight against cancer in a regular column on the site. Her writings immediately made a difference in others' lives, and the column quickly became one of the most popular items on the site, day after day getting the most "hits." What attracted the readers to her stories was the honesty in her writings. There was, at times, a little too much honesty in the stories for me as her physician. (I didn't necessarily need to know the outcomes of her dates!) But, as a friend, I was glad to better understand the inner strength of this remarkable young woman. We weren't just reading about Rebecca's experiences with cancer; we were reading about Rebecca. She put her whole self out there for us to share her experiences, her feelings, her ups, and her downs.

Cancer Happens is a guide to courage. Rebecca describes not just the cancer experience but her experience of living through, amidst, and despite the cancer. I was not Rebecca's treating physician for her treatment phase but certainly understand what Rebecca describes in fine detail. The issues that she has faced are commonly faced by not only many cancer patients, but also their family, friends, and physicians. I learned a lot from reading *Cancer Happens* and spending time with Rebecca. As an oncologist, I can describe in detail the molecular biology of B-cell non-Hodgkin's lymphoma, the cancer that Rebecca battled, but I could only imagine the inner thoughts, emotions, and feelings of the cancer patients I treat. *Cancer Happens* allowed me to experience the cancer and its treatments through Rebecca, and I am grateful for that.

Rebecca is the model of the modern patient. She has empowered herself by learning about her disease and its biology along with the disease treatments and their ramifications. This empowerment gave Rebecca confidence in her decisions and helped her to reduce stress, as she understood what was happening during the treatment and follow-up phase of her disease.

Rebecca is not out of the woods. The original cancer may return, and she is at increased risk for a second cancer (ironically, because of the treatment which helped beat her last cancer), but undoubtedly, this courageous woman will continue to beat all the odds and keep all of us on our toes.

David B. Agus, M.D.
August 2002
Los Angeles, California

acknowledgments

I sincerely thank everyone who participated in my treatment and recovery for their role in the writing of these stories—a list that's too long to include here. However, I would like to specifically acknowledge a few people directly involved in making sure this book happened: Joelle Delbourgo and Noemi Taylor for your confidence in my voice; Dr. David Agus and the staff at oncology.com for your initial belief that the unflinching truth is what the cancer community needs and wants to hear; Nancy, Scott, Heather, Melissa, Paulie, Elita, Nikki, Peggy, and John for your critiques and confidence-boosting encouragement. Thanks also to Paulie, John, Sarah, Mindy, my family, and many others who were brave enough to let me offer some of their secrets along with my own.

Most importantly, I'd like to thank my wonderful husband Larry, my most trusted and insightful critic, who encouraged me to dig deeper than I ever wanted to go.

introduction

still life #1

Perhaps it was the perfect roundness of
the fruit in the still-life print on the examining room wall in front of
me, or maybe the unbearably apologetic tone my doctor was
using, but I was having trouble concentrating on exactly what he
was saying. He pointed to a softball-sized white spot in my chest
on the x-ray hanging next to the painting and said I had cancer.
It was non-Hodgkin's lymphoma, he explained, a disease of the
lymph nodes, "sort of related to leukemia." I'd come to him three
months earlier complaining of back pain, thinking I'd pulled a muscle doing step aerobics, and this was the outcome.

Everyone I've talked to who's been through some sort of diagnosis announcement describes a different image or reaction—
falling off a cliff, a deer staring into headlights. I conjured a rapidly approaching freight train. I was strapped to train tracks watching the maniacal conductor straight out of the silent film era, red-
faced and cackling, hanging out the side of the engine, ordering
his lackeys over the roar of the engine to "Shovel faster, *faster!*"

I grabbed the sides of the examining table to brace for impact, but snapped back into the examining room just in time.

Really, like I was going to hear anything after the introduction of the "c-word." I was only twenty-two years old. But the doctor droned on for a while. I'm sure his words were very important and compassionate, but I have to call it "droning" because it sounded like someone yelling into pillows to me. I think he stopped when he realized I was staring blankly somewhere between the grapefruits in the painting and the grapefruit in my chest.

He just gave me the name of a good oncologist and said he'd already made an appointment for the next morning, pausing before suggesting I take someone with me. I'd brushed off the concerned looks from my colleagues as I left work that night and had come to his office alone, never suspecting I might need anything beyond my insurance card. To my office colleagues, my only friends in a city I'd moved to five months earlier, I giggled nervously as I told the lunch table about my appointment the week before when the doctor insisted I get some chest x-rays "just to rule some things out." Upon the doctor's request, I waited for the films outside the technician's room so I could walk them back up to the doctor's office right away. When the radiologist brought the films out to me, she looked me straight in the eye and wished me luck. When I told the story to my lunch companions, I laughed at the radiologist's seriousness, and I watched the lunch table faces go white.

Even after the CT scan the next day, I told myself these were all just precautionary measures. It was just back pain, not an aneurysm, I said. They didn't realize that I was the healthy one in the family. Unlike my health-problem- and chronic-depression-plagued sister and father, I traditionally had nothing wrong with me except a few bouts with strep throat and the memory of an unfortunate lawn mower accident when I was fourteen from which my left hand had completely recovered. I wasn't allowed to get sick and I didn't want to join them on the "unhealthy" side of the family and leave my mother—with a history of family longevity and a grandmother who nearly made it to one hundred—alone on the relatively healthy side.

But then my doctor told me the awful truth, and I couldn't go back. I officially knew that something was very wrong. Then, he looked me in the eyes with that wistful look you hope you never get from your doctor, squeezed my shoulder, and said, "The next year or so is going to be very difficult." I nearly swallowed my tongue.

I knew he was right, even though I had no clue yet what would be involved. What my doctor didn't say was worse, but I seemed to know it instinctively, the way I imagine anyone facing a health crisis does. He didn't say that the next couple of years would test my strength, choices, maturity and relationships beyond any other experience, or that they would change my life. He couldn't even tell me if I'd have a life when it was over.

As I drove out of the parking lot, all I could think about was how much I didn't want to tell anyone, partly because it would mean it was really true, and partly because I knew people would start treating me differently. So, I vowed to handle it by myself— go about my rather normal life explaining my occasional disappearances for treatments with stories of joining the Army Reserves, and the wig as a Dolly Parton-esque fashion choice. Secluded huts in the woods sprang to mind.

My vow lasted about an hour and a half. The phone rang, and as soon as I heard a friendly voice, my friend Sean's mother calling to inquire about the test results, every devious plan I'd concocted, every vow I'd made to tough it out by myself was gone. I just blurted out, "Can you believe it? I have cancer," and cried. The rest of the night the phone never stopped ringing, and secretly I was terrified it might.

The next morning I didn't go to the oncologist alone. I pretended I was fine to go alone, but about a dozen people volunteered to go with me, so I was obligated to say yes to someone. Much to my usual dismay, each treatment, check-up, and hospital stay was a team effort—family, friends, doctors, surgeons, nurses and nurse practitioners. I hated it at first. Eventually, I was grateful to have the other people to rely upon, even using them too much at times, and I learned not to resent the fact that I needed to.

This book doesn't just tell my story. It delves into themes and experiences one by one. It's about not only my struggles with cancer but also my struggle with trying to figure out who I am as an adult with and without cancer, and never quite knowing which is which. It's about seeing who I really am through the magnifying glasses of duress and pain, and then being able to live with myself after I've seen it. It's about getting through cancer with and without others, and how I treated people in both situations. It's about not knowing if I am who I would have been anyway, or whether a path and a spirit were forced like tough love upon me against my will. It's about getting through cancer in spite of what it's done and continues to threaten to do to my body and my life.

It's about getting through cancer despite it.

chapter one

cheating death

How often do you think about your own mortality? Chances are, the older you are, or the more crises you've been through, the higher the frequency. The tragically numerous and heartbreaking deaths that occurred on September 11, 2001, caused us all to contemplate it to some extent. But when was the last time you truly had to face it, consider the possibility, and discover how afraid or sad you really were at the thought?

Before I was twenty-one years old, I hadn't thought about it much at all. I'd dealt with grandparent deaths, and pet deaths, and occasionally reacted too dramatically to a death of a movie character. I was a teenager during the start of the AIDS crisis, which brought the brief thought to my mind a bit more often than was comfortable, but I didn't truly face the idea of my own mortality until a month before my twenty-second birthday, only seven months before I was diagnosed with non-Hodgkin's lymphoma. I would recall this event over and over throughout my fight against cancer.

It was the July following my graduation from college when I visited my good college friend Nick in Connecticut. Nick was my crowd's "wild guy," known to take a few crazy risks in his day— usually rebel-without-a-cause stunts like jumping out of his third-story dorm room window onto an old mattress. Or running through his dorm punching out all the lightbulbs because the resident assistant asked him to quiet down. Or burning his arm with a cigarette because I said I knew he wasn't really a nutcase.

I used to watch him with terror and admiration—part of me fearful and skeptical and part of me eager to rebel alongside him. I let him talk me into a few risky jumps of my own and still don't remember what happened the night he poured me five shots of hot cinnamon Schnapps.

He used to say that at the rate he was going, he'd never live past twenty-seven. And if he did, he'd kill himself because it was all downhill from there. It was a big joke among our friends, but a part of me always wondered how sincerely it was meant—never knowing if he really wasn't afraid of death and pain, or if he was so afraid that he forced himself to joke about it constantly.

Despite all this, Nick and I were good friends. It was more because of what we didn't talk about than what we did. We walked home from classes—many of which we took together— with plenty to say, but always choosing our conversations carefully. School, ethics, religion, philosophy, politics, writing, pop culture, drinking, eating, stories about Nick's rebellious stunts, and people and professors we didn't like were accepted and comfortable topics. Relationships, dating, conflict, sex, family, death, sickness, and people we disagreed about were not. We both had family members who were very sick at the time, and we rarely mentioned them.

We were kindred spirits in that we both essentially disliked college—the one we chose anyway. We both discovered too late that we were liberal arts minds trapped in an extremely competitive and practical journalism program. We just wanted to write and talk about the world and figure out what we thought about it all while our classmates were scrambling to the front of the classroom and kicking us out of the way for the best internships. This was the most frequently discussed day-to-day topic of conversation, and often the most practical.

We always knew when one of the taboo topics was bothering the other, but never broached it, choosing instead to provide a rowdy distraction with a heated debate over prisoners' rights or rap music. We encouraged each other's writing and nurtured a private but mutual competition regarding it. We ruthlessly criticized our professors and fellow students until our egos recovered as much as possible from their criticism.

We dragged each other through the war and toward graduation by simply being available to each other, by knowing we could enjoy each other during class even if we hated that we were there instead of at the local pub complaining about it. On the last day we saw each other at college, we joked about Nick's death wish and decided we'd see each other in Hell someday. It was a necessary but contrived finality. We intended to see each other after college, but felt we needed to make a pact in case something happened to one of us in the meantime.

Following graduation, Nick chose to recover by taking a job—in a rural part of Connecticut with dense forests and long, quiet lakes—that had nothing to do with his degree. One evening during a brief visit more than a year later, we went canoeing on one of those lakes.

We paddled out to the middle where there weren't many boats and we could see down the riverlike lake more easily. The trees were thick like the air, and any people or boats we could see could only barely be heard with the humidity sucking up the sound between us.

As we quietly paddled, I teased Nick about his approaching twenty-fourth birthday, only three years from his twenty-seventh. Out in the middle of the lake we were joking about getting old— older and wiser with responsibilities and jobs, he with a serious girlfriend and no recent injuries. Maybe we'd both cheat death and live to be ninety-five despite it all.

"There goes our good time in Hell," I said.

"I think we've already been to Hell," he said. You could see the cigarette burn scars fading on Nick's arm and my aversion to cinnamon was beginning to diminish, but we hid well that our knees were still wobbly from the broken hearts and trampled egos of those years. We were like war buddies who never talked about the final battle.

We both stopped laughing and paddling. It was suddenly wonderfully peaceful, and we drifted toward shore. We heard a rustling from the woods and waited long enough to see a soft young doe come padding out to the water, looking for a spot to drink. The woods behind her were dark and eerie in the fading light, but the water leading up to her was sparkling orange from the setting sun behind us. It was an eerily perfect moment, and I half expected to find I was standing in my grandparents' garage looking at a paint-by-numbers piece hanging next to my grandfather's rifle case. Unoriginal or not, it was beautiful.

The doe eventually looked up, turned around, and padded back into the forest, and we slowly paddled away from land to turn around. I heard a boat in the distance but didn't turn around to see—most of the lake was empty and quiet. The boat motor was getting louder so I looked up to see a speedboat, still fairly far away, steered in our direction and coming quickly. I kept paddling, thinking the driver just hadn't seen us yet. After a few seconds, the boat was much closer, and I turned to see Nick also looking at it. We stopped paddling.

"Nick, I don't think he sees us," I said. Nick didn't say anything. There was no way we could paddle fast enough to get out of the speedboat's path. It was still coming toward us and not slowing down. "Nick?!" He stood up and the canoe rocked. "Hey, asshole!" he yelled, and waved his arms. I waved my arms too. The boat didn't veer, didn't slow down, and was headed straight for us. Then it was only about three hundred feet away, and we had to make a choice: bail and hope the overturned canoe would protect us from the blades of the boat's motor, or try to get his attention one last time and hope he'd veer fast enough to avoid a collision.

Maybe it was the risk takers or the war buddies in us, or maybe the remarkably rational desire to try to prevent a catastrophic event rather than deal with the result, but we both desperately started screaming, "Heyyyy!!!" My voice cracked and almost hit a scream pitch as I yelled with desperation. At this point I could see the driver, a beer in his hand, looking toward the sunset, driving himself straight into a canoe carrying two people, and the lake shore directly behind them. About fifty feet away he suddenly looked in our direction. His eyes went wide, he dropped his beer and used both hands to turn the boat away from shore and away from us. The canoe rocked, and we grabbed the sides to steady it. The spray from the speedboat hit us, and it raced off into the sunset.

Winding through the back roads of Connecticut, we finally talked about our nearly fatal experience. Ironically, we both focused on the fact that Nick had managed to survive, which probably meant he'd live well past twenty-seven. We returned to what we knew and joked that if we had died on the lake, at least we would have died together and were guaranteed to have a merry old time in Hell. We laughed and laughed—laughing in the face of death, laughing because we really didn't want to talk about how scared we were, and didn't want to admit that we could see the fear in each other's eyes. Alone in bed that night I cried and faced my own mortality for the first time.

Seven months later I was diagnosed with non-Hodgkin's lymphoma.

Every time I contemplated the possibility of death, even for an instant—sitting in the leather chair in the hospital hooked up to various chemicals, slowly moving through the spinning CT scan, waiting for the results, searching for that concerned hand squeeze from the nurse revealing what the doctor hadn't told me yet—I conjured the image of the speeding boat coming straight for Nick and me on that lake. For a while, I was trying to figure out if, honestly, I would have minded if I'd died that day. Fear of death was never something I'd considered real—just an animal instinct that forces an adrenaline rush to protect oneself. I was trying to figure out if I was prepared to move on to the next life. Honesty forced me to concede that not only was I not done with this one, I was desperate enough to hang onto it that I surrendered to the frightened, primal scream that desperation forces.

Eventually, the incident gave me a strange sense of comfort. I'd escaped death at least once already, so I could do it again. At times the thought crept in that it might have been my ninth life—the first eight used up jumping off of things with Nick. But I chose not to dwell on the possibility too long.

Ironically, Nick was one of the last friends I told about the cancer. I knew it would be like getting injured on the battlefield and then crying out to your buddy to help you when he's still busy fighting. It seemed . . . inconsiderate.

But with hindsight I know I didn't want him to hear my voice—the voice of his supposedly fearless buddy. I wasn't sure how well I could hide what I was really going through. I waited a couple of months into treatments, having already mastered breaking the news while inserting many hopeful phrases and projecting an appropriately positive attitude—what I'd learned were the easiest words for people to hear, exactly what they wanted to hear. I told him over the phone, and I didn't take a breath until I'd finished the whole spiel of explanation and carefully chosen details.

He didn't say anything for a long time, so I continued saying cautiously optimistic words and knew I'd just keep going until he said something. Finally, he stopped me and said, "I hear what you're saying, Bec. Everything's going okay, good prognosis and all that. But do you realize how scared you sound? You sound so fucking scared." His voice broke then, and I knew I'd broken him. The Fearless Ones, the tough-as-nails duo that had dragged themselves through the fires of young adulthood, were frightened of death. He'd seen it. He'd seen what I didn't want him to. I was scared and weakened—struck by a deadly disease when my body was supposed to be at its strongest. He'd heard somewhere in my voice that same squeaky desperation as my screams on the lake. Only this time he couldn't stand up and wave his arms and prevent catastrophe, and we weren't sure if I'd be around to laugh about it after the fact.

I'm sure it didn't escape him either that he was the one who was supposed to die before he was twenty-seven and now was struck head-on with the fear of death—either a very familiar state or a foreign one—all because of me. We got off the phone and didn't speak for a long time.

At night I often dreamt about Nick and me on the lake just as I fell asleep. As long as the speedboat continued to turn away from us, to veer away quickly in my mind, or as long as I woke up before the blades reached me, the longer I could muster the strength to fight off the disease. Because of this, whether Nick was actually beside me or not, whether I ever admitted whether I honestly wanted him there, he was a part of my fight against cancer. He faced the fears, joys, and failures of becoming an adult with me and helped me become who I was, who I needed to be, at diagnosis. More than anything, he faced death alongside me on that lake—the first time I had to face my own mortality.

I clung to both the dream and the memory. They reminded me of a lesson both Nick and I learned together: It's possible to cheat death, but I should be careful about laughing so unflinchingly in its face just because I'm young. It eventually will bite you in the ass just to prove itself truly worthy of every living being's instinctive respect. It also showed me that no matter how brazen or frightened or sad I may have been while fighting cancer, or may yet be in life, I'm not ready to jump ship.

By the way, Nick turned thirty-three in January. He's doing very well.

chapter two

normal

Normal is a relative term. For Americans it's everything from 2.4 kids and an SUV in the driveway to begging for change on subways to buy a hot dog. Any life situation has its obstacles and joys, but almost everyone's idea of normalcy conjures images of stability and predictability, even for those who claim otherwise. We all need the cocoon to return to when the jolting blows hit, and we all know they eventually will.

The day I first went to my oncologist was the day I started to lose my somewhat unappreciated sense of normalcy in my life. I had what I thought was a fairly normal life before that. Most of my time was spent at my nine-to-five-plus entry-level position at a local public relations firm. Most of my off-work hours were spent in a tiny apartment in a four-unit building with my stray cat, Lucifer, who lived up to his name with others but adored me. My place was decorated with photos of college friends, framed art prints, and hand-me-down "retro" furniture in the living room, covered with throws from a home store catalog. I had plans to buy a stereo and a good set of kitchen knives. I couldn't yet afford a gym membership on my small salary, so I did step aerobics to a videotape, on an area rug covering my hardwood floors. The eccentric cabinetmaker with a beard—and a marijuana habit you could smell the moment you entered the building—living in the unit below me didn't mind the forty-five minutes of stepping. He told me so when he'd periodically stop by and ask if I wanted to go to his place for a smoke. But the mysterious "security guard"—I always suspected he was in the CIA—who only worked at night and frequently went away for long trips—living in the unit next to mine came over every few weeks to complain about the rhythmic "pounding." We finally agreed that I'd put towels underneath the rug when I exercised, and I never saw him again.

On the weekends, I went to museums, parks, and movies by myself, occasionally having friends into town or going out for drinks with coworkers. I'd been in Cincinnati for five months, not long enough to make many local friends. They began recognizing me by name at the video rental store down the street, and my phone cord was long and stretched from use, as was my phone bill.

The majority of evenings were spent in conversations with one or both of my more interesting friends, two in particular—Sean and Joanie. Neither of their lives were anything resembling normal by middle-American standards—they both were studying and pursuing a career in the theater—making mine seem utterly dull by comparison, and I continuously envied the choices they'd made.

Sean and I had known each other since grade school, and by the time I moved away, had embarked on two full-fledged love affairs, five let's-just-be-friends-but-still-sleep-together love affairs, and at least twenty failed let's-not-be-anything-at-all attempts. He was tall, dark, and handsome in an unassuming way, his Italian heritage more apparent than his Irish. People were drawn to his combination of nonthreatening, quiet openness and bad-boy sarcastic humor. He had a smile and a cunning twinkle in his large brown eyes that revealed everything from naïveté to an unconscious arrogance. An actor, a very good actor since he was Hally in *Master Harold and the Boys* at age fifteen, he always seemed to recapture my heart from whatever hyphenated phase our relationship was most likely in whenever I saw him on stage. He revealed more to the audience than he could to me, and he always seemed to be looking directly at me. I wondered if everyone else in the audience felt the same way. It was a most effective aphrodisiac.

Joanie also had known Sean and the rest of his large family —four older brothers and a younger sister—for many years, but we didn't hit it off until I graduated from high school a year earlier than the two of them and left for college. When we could have been hanging out in the same artsy cliques—we were both in the school system's performing arts program—we weren't really interested in one another. But once we added a few meaningful letters and the distancing comfort of not having to see each other every day, we were the best of friends. While I went away to college, she stayed in Dayton—at first living with her family, headed by middle-aged hippies, in a beautiful old creaky house decorated by their mother's art and the kids' art-class masks and watercolors, then in a place of her own.

Joanie wore dark oversized clothing, especially big shirts and overalls, nearly every day for the first few years of our friendship. To the unfamiliar eye she appeared unapproachable, even angry, but she was extremely likeable. She made friends easily, with her unique combination of brashness, strength, and vulnerability. She had chosen a career in the theatre, but as a stage manager and costume designer—places where she was in control of something and could wear what she wanted, but still felt part of a thing bigger and more fabulous than herself. It was the perfect career choice.

Joanie and I were friends almost by default. We enjoyed each other's humor and strong opinions about the world, spending hours in one sitting spouting off about something or someone just because we could, and we enjoyed the fact that we nearly always agreed, helping validate our sometimes unpopular opinions—like that most people our age were uproariously stupid, and we were two of the few people who knew anything about anything. God, we were young. Joanie also helped me quietly celebrate my twenty-first birthday—a between-college-semesters birthday for me—with an unassuming bottle of wine, a couple of movies, and three hours of sitting on her bed talking about the status of my relationship with Sean, her latest production, and, finally, where we both wanted to be at my twenty-second birthday. After college I shared a house with her for a couple of months while I job-hunted. Then I chose to move to another city to pursue a "professional" career while she became immersed in the local arts scene and a new boyfriend. After a while we seemed only to be linked by common friends—Sean being the most prominent— and a long history, so we remained in each other's lives.

There were Sean and Joanie, over there, with their interesting artistic lives, and here I was in my normal and, let's face it, decidedly boring existence in Cincinnati. Nevertheless, I was convinced I was happily free as could be. My obsessively independent nature told me I could go skinny-dipping or join the theater or travel to Italy at any time, and only I decided where the limits were. These free-spirited twenty-something-like freedoms I believed my friends possessed probably did exist for me, though who knows if I ever would or could have taken advantage of them. Still, I found comfort in their existence, but only as long as I had my snug apartment and stable job to come home to.

Then my body decided that options were evil and began eliminating them systematically.

The day after my diagnosis, a Thursday and the day of my first oncology appointment, I became the proverbial pincushion. I didn't just feel like it, look like it, or act like it; I was an actual pincushion—a place for healthcare professionals not only to stick sharp objects, but also to store them for later.

First, various fluids I'm typically trying to keep to myself—blood, urine, stool—were extracted at record pace and then passed around to several different people who wanted to look at them. At one point I lay on my stomach so the doctor could grind into my hips what felt like a corkscrew to extract bone marrow for a biopsy. I watched the nurse who was holding my hand reach up and wipe the sweat, earned while combating my hard young bones, from the doctor's brow at least six times during the twenty-minute procedure.

It continued the next morning when the doctor inserted into my chest a port-o-catheter, a round plastic disk that sat just under the skin in my chest, attached to a tube going into a main vein. It offered a "more comfortable"—we wouldn't want the poison to be uncomfortable would we?—direct line to my bloodstream. The alternatives were to access my arm or hand with every treatment, a practice that is now used only under unusual circumstances, or to insert a tube catheter that would stick straight out of the skin in my chest and swing around underneath my clothing between treatments. They didn't offer me a choice, really, but I knew the less conspicuous disk would be the best option.

Though I had more tests and my first treatment the following week, the nurses decided to insert all the necessary equipment right then, saying it would make the tests go more smoothly. Apparently, only some of the hospital personnel knew how to access such a catheter, and they went into a big story about how one of them would have to be called and they'd have to go all the way there to insert it themselves and on and on. I wasn't sure how I felt about having a device inside me that the majority of the healthcare community didn't know how to use, but my almost desperate trustfulness didn't permit me to argue with them. Before I could blink, a nurse was coming at me with this crazy-looking L-shaped needle with a syringe attached. She said quickly, almost after the fact, "Little stick," and then put a good amount of her weight into pushing the short bent end of the contraption fairly far into the disk in my chest before taping down the whole thing, now dangling from my chest like a tetherball. She told me that sometimes such a wound bleeds a little, but it would just look like my chest sprang a tiny leak. "This seepage is normal," she said. "Only call us if it looks alarming." I tried to imagine what "alarming seepage" looked like and couldn't imagine any kind of leak originating from my chest seeming unalarming.

Wrapped up like a wounded soldier, I was allowed to go home and resume my normal life until Monday. I put on heavy, high-necked clothing and went back to work that afternoon, said hello to the pothead when he peeked his head at me as I returned home, and tried to ignore that everything was not-so-slowly changing.

That weekend Sean and Joanie planned

a last-minute visit, certainly motivated by a need for comfort and escape from the fast changes and growing fears we all were feeling. Sean had been visibly unnerved by the news, and at first I kept thinking how young we all were to seriously be contemplating my fate. The night of my diagnosis Sean and I were in an undetermined phase—as yet untitled, you might say. Nevertheless, he was so distressed he stayed with me that night and accompanied my father and me to the oncologist the next morning. As I attempted sleep, he'd held me tight for hours as I ground my teeth and twitched in my anxiety-riddled nightmares.

At one point—I'm not sure if he knew I was awake or not—he squeezed me tighter and whispered through what might have been tears, "Don't worry, Bec. I'm not going to let anything happen to you." The warmth of his body was suddenly very comforting, and I felt my body relaxing. He'd chosen to pick up some of the weights on my back, and I was physically lightened from it. I'll never forget that.

Similarly, once I was diagnosed with cancer, and despite the good prognosis, Joanie became notably frightened and went out of her way to be a comfort to me—calling almost daily and scheduling and promising visits. Who knows if we would've remained as close, or if she would have let me depend on her so much, if I hadn't gotten sick. But we didn't think about that because we knew things were different now.

On this day we all were bosom buddies, pretending that everything was "normal" because we simply, and desperately, needed to be with each other.

So, the three of us decided to do a little shopping, order some dinner, maybe rent a chick flick against Sean's protests, and watch it in my cozy little apartment—a very normal Saturday night for me. They said they just wanted to spend a weekend like "old times," but I couldn't remember the last time we went shopping together. I knew they were using the sterility of the mall and the lure of take-out Chinese to keep our minds off the obnoxiously large syringe sticking into my body and what it was there to do.

Upon Joanie's insistence, she and I, both tomboys at heart, headed for one of our favorite escapes. A few years earlier we had bonded over a shared and, to us, embarrassing passion for silky lingerie. What better way to forget your troubles than to go to that place of synthetic romance and figure-flattering mirrors— where it's okay to imagine you're on the cover of a romance novel and not giggle to yourself? We'd used this setting before to forget our troubles and escape the crisis at hand, and we knew it worked like a charm. So, we left Sean, who was on the prowl for a new jacket, and headed toward the lingerie store dressing rooms, silky nighties in hand.

We chatted with each other, discussing what colors we were wearing these days; Joanie had abandoned her traditional black for lighter colors, and I was beginning to wear more black. Still, I had chosen a short blue number for the dressing room and she had chosen a long black negligee and a pink nightgown. Old habits die hard, we said, and laughed at ourselves.

I slipped the slip off the padded hanger and over my body in one practiced movement and remembered why we'd used this as a great escape. The slip settled on my shoulders perfectly like I was gliding through soft, warm water. It was that scandalous, luxurious feeling of simultaneously wearing clothes and being nude.

The first time I'd been given such a piece of lingerie I'd resisted it, believing I wasn't girly or my barely-B cups weren't voluptuous enough. The source was a handsome but dim college upperclassman I'd had a secret love affair with for a month during my freshman year before we both lost interest in each other—how very "lingerie store" of me. (Obviously, Sean and I were on a "break.") My upperclassman wasn't a man known for his generosity, but also wasn't undiscriminating when it came to women. He ignored my resistance and finally put the silky slip on my body for me. He looked at me encouragingly, but I wasn't even paying attention. I had to look at myself in the mirror to realize I had anything on at all, and then couldn't believe what I saw. I was soft, but not weak. I was feminine, but my suitor's eyes let me know I was very much in control of the room. It made my hard edges smooth and my big butt "shapely." It was a feminine high I felt every time I wore such an article after that.

After letting the nightie rain down over me, I looked in the dressing room mirror and my sense of escape disappeared.

Instead of a soft, womanly tomboy in blue silk, all I saw was a white bandage covering the top of my left breast with a syringe sticking out of it, looking like the damaged remains of a long night of drug use and making the blue softness of the slip bulge awkwardly. There was a little brownish something seeping from the entry point of the needle and I jumped slightly in surprise and disgust. Then I remembered that "seepage was normal." But what was normal anymore? Normal was no more.

Joanie started chatting about how she wasn't sure if her boyfriend would like hers, so she probably wouldn't buy it. How did mine look, she asked, as she looked up at the mirror. I wanted to say I looked like a prostitute who had been stabbed on the job, and I felt like I was going to throw up. Instead, tears came to my eyes as Joanie turned to hug me, her eyes also filling with tears. Then I took the slip off, placed it delicately back on the soft hanger, and covered up the syringe with my high-necked clothing.

I left the dressing room and the drama I'd played out inside, but I had to buy the nightie. I put on my soon-to-be-automatic positive outlook face—"Remember, the prognosis is good!"—and proclaimed it would "look fabulous with a hospital gown." I smiled at them, and they hesitantly smiled back as I raced to the counter and purchased my goods, pausing to offer a couple of "I really shouldn't"s and acting grateful for their "You deserve it"s.

I walked out of the store and out of the mall, promising myself not to return until I could live up to the normalcy I was so desperately forcing with every smile muscle and every playful swing of my shopping bag—until I could walk in and know that the silk and mirrored illusions could actually help me escape from whatever was ailing me, or at least help me get back to anything resembling a normal life. For the first time, I knew that might never happen.

chapter three

a steroid buzz

Anti-inflammatory steroids are used to treat many illnesses, including cancer, as part of some chemotherapy regimens. Possible side effects include increased blood pressure, sudden weight gain, insomnia, nausea, vomiting or stomach upset, fatigue or dizziness, muscle weakness or joint pain, increased hunger or thirst.

Translation: Every three weeks, I hopped on a five-day physical and emotional roller coaster ride called chemotherapy that included high doses of these steroids. These are some of the most memorable days of my experience:

FRIDAY—DAY ONE, ROUND TWO OF CHEMO—1:22 P.M.

I'm sitting in this big plastic easy chair that sounds like a rain slicker rubbing against itself, or a bodily function we try not to mention in mixed company, every time I move. Just waiting for them to hook me up. I've been sitting here for a while, actually, but every so often the outpatient cancer care nurse pops her head in and checks on me. "How are you doing?" It's such a loaded question, and at the moment she truthfully is only interested in whether my tubes are leaking, so I just say, "Fine." A half hour or so ago she tried to get me to take my hair out of my hat so she could see how much had fallen out. She didn't even ask; she just started taking it off like it was a medical necessity to see my hair follicles. My reflex reaction was to move my head so she couldn't get it off. Even with my instinctive protest, she tried again after stopping a moment—sort of like trying to trick a puppy into giving you the toy by pretending you don't want it anymore. But I reacted instinctively again and didn't apologize, didn't say anything. I couldn't look her in the face and couldn't imagine some stranger poking around on my patchy scalp when I was feeling so ugly. She stopped trying. I feel badly, since she's a very nice nurse and she truly means well. She's probably seen a lot of very sick people die. A little patchy scalp of a twenty-two-year-old with a good prognosis is nothing to her.

I'm staring at the television and switching around looking for something other than talk shows with bold quoted headlines like "Mom, your other personality dresses like a whore!" and "Honey, I'm carrying your son's twins for the third time!" There's got to be something more challenging on for those of us stuck in an easy chair on a Friday afternoon dreading bags of poison that hopefully will save our lives. It seems like there should be something more promising to choose from, something telling different, more hopeful kinds of truths about humanity—something so I can maintain my appreciation for the world I'm trying so desperately to remain an active participant in. At this moment, I'm having to dig deep.

I can think of at least a hundred places I'd rather be sitting than here. What a wonderful way to pass the time. Let's count them: 1) beneath a tree reading; 2) on a train going across the Serengeti; 3) on a plane flying over the Grand Canyon; 4) in the fourth row watching a great play, preferably starring Sean; 5) in a sidewalk café with friends on a cool and sunny autumn day; 6) on my couch petting my cat Lucifer while watching *West Side Story;* 7) at a high-priced martini bar with my legs crossed outside the slit in my skirt, waiting for my handsome date to check my coat; 8) alone in a bar in a seedy part of town, nursing a beer and watching the bartender break up a fight; 9) watching a documentary about tree surgery; 9) handcuffed to a chair at a police station; 10) naked in front of an art class filled with prison inmates. This can't be going anywhere much better than what is on the television above me, so I should stop.

Part of me just wants to think of anything else besides my conversation with Sean last night. As ever, we remain undefined and categorized only by what we can and can't say to each other comfortably. We can say we love each other, that we desire each other, that we'll always be there for each other. These words, said by a man who's been an enduring friend over many years, bring immeasurable comfort and almost physical warmth to me when I hear them. They come from a man who surprised me in a suit at my sophomore year piano recital and coached me through my first and only acting role in our high school's production of *The Mikado.* Who resented it when I didn't tell him when I thought I might be pregnant with his baby. Who took every panicked, stressed-out phone call I made during college, and never asked me to do the same but was grateful when I did. Who raced to the intensive care unit with me when my sister jumped off a third-story library balcony. Who shares his family's Christmas with me nearly every year and sits for hours at the dining room table drinking coffee with my parents discussing their collection of antique glass. The man I would do anything for.

Despite our relationship—perhaps because of it—more often than I enjoy, that little voice on my shoulder points out the elephant in the relationship: There is a limit. We're not allowed to have any promises, any commitments of any kind. We must remain this vague definition of friends anchored by the oft-spoken phrase, "I'll always be there for you." Is it wrong to need to know more? In what way will you be there for me? How often will you be there for me? How much am I allowed to ask? I've started referring to him in my head, and sometimes to Joanie, as my half-boyfriend. It's like having a half-brother—he technically is a brother, but with a qualifier of inherent distance and clearly labeled distinction from a full-fledged one.

Really, I think what's bothering me is that I had these fantasies of that postdiagnosis "moment of clarity" that some survivors speak about. That profound truth that, once it's actually spoken and accepted as reality, makes all the difference to both you and that special someone. Rules, neuroses, and baggage all fade into things that just don't matter anymore now that everything has changed. I fantasized about that moment with Sean, wishing for the only good thing I could think of coming out of this cancer thing. Maybe I am too insistent about repeating the good prognosis and how "everything was going to be fine after a few treatments," but that moment of clarity happened for me alone, not for him. That pisses me off.

Last night, the night before my chemotherapy treatment, a night when I was distracted and worried about what I was doing to my body and soul, I needed for all of this to be as serious to someone else as it was to me. I needed for it to make all the difference to someone else too. I told him, "I'm scared, and I need to know that someone else is just as scared about all of this." Sean was silent for a moment, perhaps resisting the urge to say what he really felt, but then he went back to the established and previously acceptable phrase that was only half as comforting: "I love you, Bec. You know I'll always be there for you."

It was very sweet and, I am positive, absolutely sincere. But it didn't work.

FRIDAY—DAY ONE—1:49 P.M.

My mother's still out looking for juice in the kitchen. I peeked out the door and there's a young guy, even younger than me, sitting across the hall with a baseball cap like mine over his bald head, staring blankly at the TV. Is that what I look like? His eyes are tired, his IV is dripping steadily, and he seems to be used to the drill. His mother is talking quietly with his doctor outside the room, and I can't tell if he realizes it or not. He probably does, but he may be too exhausted to care anymore.

Okay, let's think about this. Prepare. First the saline fluids, then the antinausea drugs, then three bags of the high-powered stuff, one of which is bright red. Then they send me home with a load of pills to take for exactly five days. The pills make me a little jumpy. Where's Mom?

FRIDAY—DAY ONE STILL—3:15 P.M.

I'm hooked up to the second bag of chemo drugs now, and I can't believe I'm hungry. I have a needle sticking out of my chest and my mouth tastes like I swallowed acne cream, but I'm craving chicken . . . and roasted potatoes. My mother is sitting here with me reading, and I'm still flipping around the television—three seconds per station now, as opposed to the second and a half I was spending on each ten minutes ago. This probably would be a good time for us to have one of our mother-daughter blab sessions—where we talk about my friends and relationships, gossip about my colleagues, and tell stories about the quirks and hilarity of my mother's kindergarten classes.

You haven't seen my mother at her best until you've seen her teach. Nearly thirty years of experience certainly help, but some teachers are just naturals with little kids. My mother possesses that unusual and perfect combination of softness and angles, compassion and rigidity, structure and creativity, warm smiles and well-placed scowls, to be the perfect kindergarten teacher. She commands a room of difficult five- and six-year-olds like a five-star general and comforts them like Mother Hubbard.

Some of the proudest moments I ever had were when I helped her in her classroom for a day during Spring Break. I was the oddity of the day and a great way to get the next Letter Person doll blown up, but my mother was the star of the room. A woman who normally has trouble getting through a movie where someone dies without audibly blubbering became the focus of every child's unconditional respect. She hadn't simply asked for this respect, she'd earned it through consistency and a patient ear.

The hundreds of sad and fantastical stories we'd always heard her tell in enthusiastic detail over dinner—like the time a mother threatened her with a knife and the year she had a student named Placenta—told us that she loved her job, but this proved she was exceptional at it. Still, I love to hear those stories. But not today, and she doesn't seem in the mood either.

FRIDAY—DAY ONE STILL—3:49 P.M.

I'm starting to calm down now after the initial rush of being poisoned. Within seconds of the nurse hooking up the first bag, my heart started to race, and I could smell the chemo drugs through my skin. My body seemed to be rejecting it—speeding up its processes and switching into full alert the second the foreign fluids entered. It was like all the parts in me were waking up, sitting straight up in bed, and wondering what the hell I was doing to them. They all simultaneously began screaming softly from their little corners of my body. It's taken me ten minutes just to get my breathing down to normal without alarming anyone.

Now she's hooking up the red stuff—what the doctor calls the "hair killer." Even the first time they gave me this stuff I could feel it actually killing my hair. It felt like when you have a fever, how all your joints and tiny muscles ache, but more than anything you can feel every hair on your head and neck and arms because they hurt every time you move. That's what it feels like on the top of my head right now as I watch the red drops drip and the steady flow of beautifully colorful toxic liquid going directly into a vein close to my heart and speeding directly up to the precious few hairs still protecting my head from nakedness and cold. No wonder everyone's screaming in there. I desperately want to scream along with them, but I don't think it'd do much good.

I have to go to the bathroom, but I don't want to negotiate all of the wires to roll down the hall. And my pee is a scary dark yellow, almost orange, after this red stuff. And the needle pinches when I move too much. I don't want to risk any needles poking through any veins and poison spilling out into an uncontrolled spin in my body, attacking whatever it lands upon. But I guess I can't imagine that would feel much different from what it feels like now.

SATURDAY—DAY TWO—2:52 A.M.

I haven't slept yet. Mom is sleeping on the foldout in the living room, but I'm wide awake. My stomach hurts, but I'm hungry. We ate the chicken I was craving tonight. I scarfed it down, hiding the six little acid-tasting pills in my macaroni and cheese and mashed potatoes, and then promptly threw half of it back up when I accidentally bit into one. Are they still working? Did I throw up what's going to save my life? Should I take more? My stomach is turning at the thought. Mom didn't make a fuss about the throwing up, and she didn't come in to hold my hair and my sides like she did when I was a little girl. But I caught her furrowing her brow and turning away to do dishes when her face started turning red. In my "cozy" apartment, my utilitarian kitchen is actually in my living room, so she can't rush to it as a refuge when something's bothering her like she does at home. She probably hates that, and I feel badly for not having a guest bedroom for her to cry in privacy.

My racing heart feels like it's actually making the skin over my chest bump with each beat. It wants to go run a mile, but my throbbing legs are screaming. I think I actually heard my body screaming at me today, and I can still hear it every once in a while. Am I crazy? God, I'm tired, but my mind won't let me find the quiet, and I'm afraid of what I'd dream. My bladder won't shut up either. I'm still peeing IV fluids every fifteen minutes, even though they told me to drink lots of water. I'm thinking about missing work yesterday and what people will think if I miss it again Monday and whether my Mom was really getting juice or conspiring with my doctor. What do they say when I'm not in the room?

I have to pee again.

SATURDAY—DAY TWO STILL—10:48 A.M.

Haven't done much today except shake my foot and watch movies on cable with my mother. Not that I could do much more than that. I have this nervous energy that's growing faster, even faster than my frustration, but my body is sinking into the couch like a shot put. Even so, I probably should be more tired considering I was methodically poisoned two days ago. Wouldn't it be faster for me to drink a cup or two of bleach and be done with all this cancer crap?

Mom's being a good mom. She's quietly just being around, watching me without staring, making me feel like she's just there to comfort herself, when she knows I shouldn't be alone, encouraging me to eat and drink my water. I'm not hungry at all, but my stomach feels hollow and sore from vomiting last night. She's done the dishes and cleaned the kitchen and fed Lucifer and folded up the uncomfortable foldout couch she slept on without complaint. Mom's never this quiet—an endearing trait that I often tease her about. She'd tell the neighbor's dog sitter about every one of the collectible glass pieces she'd acquired in the last year if he'd give her the chance. Her Christmas letters would make Tolstoy feel like he had to struggle to keep up. And she's rarely hesitant to express her annoyance with something if she can help correct it—she complains to the grocery store manager if the selection of string beans or celery seems old and has no tolerance at all for anything that can be interpreted as a patronizing tone in response.

Today she's very quiet, so I know she's really worried.

SUNDAY—DAY THREE—9:59 P.M.

Mom left yesterday evening. I told her I was okay, but I really didn't enjoy seeing her go. Okay, this sounds crazy, but I can feel the poison inside me. Not only can I feel those stupid pills burning a whole through the crackers I ate and through my stomach lining, but I can sense the poison burning through my bloodstream, especially in my chest where the tumor is. It's eating it, bubbling like hydrogen peroxide on an infected cut, and it's the red stuff I can tell because it's coating and shrinking it like some thick acid syrup. It burns when I breathe in and only in that spot so I know it's working. If I drink water or move too fast, will it flush out more quickly? Maybe the blood will start to flow too fast and the syrup get too thin and the acid wash away before it can eat the whole thing. It's hungry and it's trying to save me. I'm afraid to move. I'm going to slowly put the pen down and go to sleep propped up against the wall with my head resting on what I think is a hanging photo of Cape Cod. It's the best thing.

MONDAY—DAY FOUR—11:09 P.M.

I got dizzy at work today and called Sean and cried. I didn't want to be there, and I'm sure everyone hates me or just feels sorry for me or resents me because I take days off, and I disrupt their lives. I can't pull my weight, and they know it, but they'd never say it because they feel sorry for me. I don't want to be here! Maybe I'll call Sean again.

TUESDAY—DAY FIVE—1:15 A.M.

Sean hates me now. He asked me why I was calling for the third time, waking him up from the only sleep he's gotten in days each time, and I lost it. Why am I calling? Well, I've got a little thing going on in my body right now, and I'm up thinking about it instead of asleep fighting it, which would be the healthy, sane thing to do. He just said he was really tired and had a big day the next day. He didn't hang up and asked me what was wrong, but I could hear it in his voice that he didn't want to talk, and I couldn't bear putting the slightest inkling of "pest" or general notion of "burden" into anyone's head, so I hung up. He didn't call back, but I picked up the phone more than a dozen times to call him. I didn't.

It must be hard for the people in my life—I know it is—to carry around what's happening to me as they try to function in their normal lives. It must be hell, in fact. I try every day to understand that more and more and not ask too much of them. But what happened to "I'm always there for you" and "I won't let anything bad happen to you?" Why is there a limit to what I can ask of people when there doesn't seem to be a limit to the annoyance and pain? Everyone outside of me can always get off the phone or go somewhere to escape the truth of it all when it is too much to bear. But I can never do that. It's always there, every morning when I wake up and realize that the nightmare I had about having to turn my life upside down just to keep from morphing into a ball of stinky malignant cells and pain was just a sleepy recollection of the day before.

I wouldn't consciously choose to share that kind of nagging heaviness with anyone, but sometimes, especially right now, I wish I could. I just wish someone could take a little bit of it so it wasn't so burdensome. I wish there was someone who wanted to.

TUESDAY—DAY FIVE STILL—11:43 A.M.

I cried myself to sleep last night. My eyes are puffy today, and what's left of my hair is greasy and it hurts. Last night the follicles screamed at me every time I turned my head against my pillow trying to find some peace in my mind—enough to be able to close my eyes to silence instead of manic ramblings about burdens and real-life nightmares. Maybe I should just shave all my hair off.

The only thing I want to eat today is a couple of pounds of Buffalo wings and curly fries—I can even smell them I want them so much. But how do you hide those godforsaken pills in curly fries and I think that Buffalo wings would burn if I threw them up, besides the fact that they'd make me even more bloated than I already am. I thought chemotherapy was supposed to make you lose weight. My doctor said most people don't mind the steroids because they give them energy and appetite and they feel like "a person" again. I'm happy for those people, but before I took these damned pills I was doing step aerobics six days a week and could eat an entire pizza in one sitting and I've never felt more like a freak in my life.

WEDNESDAY—DAY SIX (FIRST DAY OFF OF STEROIDS)—8:04 A.M.

Slept exactly forty-eight minutes last night, and my shoulder is killing me from lying on the floor of the bathroom for three hours—intestines hurt—moaning in the bathroom, desperate to have it just end and not willing to pick up the phone to burden anyone. Sleeping with my hand on the phone and can't seem to stop crying and aloneness is physically painful. Nothing in my closet is clean—should I call a cab or a dry cleaner?

THURSDAY—DAY SEVEN (SECOND DAY OFF OF STEROIDS)—6:08 P.M.

Work was okay today. My body's starting to recover, and I don't think people stared at me so much. I'm still not sleeping more than a couple of hours a night, and I have to have the radio on at my desk in order to keep the self-inflicted paranoid ramblings at bay. Can't seem to muster the humility to call Sean yet, but I know he's really busy with his latest production so he probably hasn't noticed I haven't called him back yet. Or he's so angry he doesn't want to talk at all. I'd like to be angry with him because then I wouldn't be so beside myself with fear that he'll never call again. Maybe I should just call and find out.

FRIDAY—DAY EIGHT (THIRD DAY OFF OF STEROIDS)—5:26 P.M.

I got some sleep last night for the first time since chemo. Sean called and, ironically, woke me up just to see how I was doing. He didn't mention the hanging-up incident at all, and I'm sure it's best to just leave it that way. He's a good friend, or half-boyfriend or whatever he is, for leaving it that way.

I'm going this weekend to get my hair shaved off so it'll stop screaming at me. Some of the folks from work are going out for Buffalo wings, and I think I'll go along.

What's in those pills anyway? Only a couple of weeks until the next treatment. I can't imagine what the next one will be like. I'm tired just thinking about it.

chapter four

hair stories

All my life I have been preoccupied with my hair. It's thick, dark, and hard to control, and not just on my head. From the day I was born, on my head grew a mass of full, thick locks that my mother's hairdresser always said was like cutting through chicken wire. It was soft as fur, but your fingers could dig for days through a forest of dense hair before finding the clean, healthy scalp underneath. As a child, I wore it long and simple, needing only baby shampoo and a few hours in the breeze to make it shine and smell like a wheat field. The rest of my family had fine hair, straight as sticks, and they were jealous that I'd inherited some rare family gene, probably from a Scottish tribesman in a muddy kilt with beautiful thick hair down to his waist.

And yes, it was just as thick, dark, and hard to cut or get rid of everywhere else as well. Of course there was lots of it, rarely appearing in the most convenient or attractive places. This didn't bother me much when it first appeared.

Then the summer between fifth and sixth grade I was at the Pine Forest Camp swimming pool. My cabin, the Frazier Firs, and the closest boy's cabin, the Pine Cones, decided to play a game of keep-away. I had just cornered Richard, a redheaded and blue-eyed young man of particular height and looks. I raised my still baby-fat chubby arms to attempt to tap the ball down from his long, pale arms, only to stop him in his tracks and cause him to stare at my underarms—better known as armpits by other eleven-year-olds.

"What's that?" he said. I looked at the reddish-brown tufts of hair forming under his own arms and wondered why he'd even noticed, considering his own sprouts. "Hair," I said, almost missing a chance to steal the ball. I started to turn to hurl it at my cabin mate, but not before catching a look of disgust running across Richard's freckled face. I heaved the ball at my bunkmate's shoulder, watched her squeal in a way only small blondes without a follicle of visible body hair could, and swam to the corner of the pool. As I let my legs float up in front of me, I could see the thick dark hairs floating toward me. I looked up my thighs and saw the dark curly ones peeking out over the edges of the suit. I was mortified, and the obsession began.

As a teenager I spent years experimenting with hairstyles and lengths, none ever being good enough. I spent way too much time curling it in the morning, brushing it during the day, and then redoing it at night. Once when I spent too much time on the bangs I was sure I needed because my forehead was too big, I got frustrated and cut a chunk of them off with pinking shears. Then I spent more hours with the curling iron and hairspray trying to figure out how to cover up my error.

I shaved my legs and underarms every day, especially when taking gym, for fear of being caught with stubble. In the summer—pool season—I shaved my bikini line so much that my sensitive skin started to rash and irritate. In gym class the red bumps on my

legs were so obvious that it didn't matter if my classmates ever
noticed the hair. They all noticed the bumps.

After every razor cut and burn I cursed the hair gods and
wished for a way to permanently remove it all. My fear was that
I would be outed as the awful hairy beast I truly was, and would
end up banished from the kingdom of normally follicled people
forever.

Yes, it was a true obsession, or at least an easily diagnosed
neurosis, but that's actually how it was, I swear. I'll give you an
example. When I was fifteen, I was on *Good Morning America*
with three other teens to interview then–Surgeon General C.
Everett Koop. In high school, Sean and I were cohosts on a
Planned Parenthood–sponsored local teen-oriented radio show,
and later its cable access television offshoot, called *Teen Rap*.
Give us a break; it was the '80s. On our terribly "controversial"
show we discussed with experts from various organizations social
topics of interest to teenagers, especially sexuality, AIDS and
other sexually transmitted diseases, teenage pregnancy, sexual
taboos, sex in the media, date rape and on and on. We occasionally talked about things outside the general area of sex like
teen suicide, school violence, even general hygiene, but mostly
we talked about what we knew was the teenage population's
favorite topic every week on the No. 1 station in Dayton, Ohio,
in the terribly sought-after timeslot of 7:30 on Sunday mornings.
The producers thought we were novel, we thought we were terribly bold, I didn't have to do my hair for radio, and each week,
without fail, we achieved our goal of making someone blush at
least once with our fearless questions. We thought we were one
step away from *60 Minutes*.

So, when *Good Morning America* called looking for teenage girls who weren't afraid to talk about sex on national television, they immediately thought of me. This was by far the most
exciting thing to happen in my young life, and my preparations

were key. I wrote thoughtful, fearless questions and discussed issues with producers during mandatory preshow phone calls. I thought through what I would say to my fellow teens in the green room to be sure we'd have a good rapport on the air. Then, I looked at myself in the mirror. Of course, my shoulder-length straight brown hair with '80s curled bangs wasn't good enough, so I made an appointment to have it permed quickly before I left.

The next couple of days were a whirlwind, but I remember every minute. My mother and I were flown to New York and transported around in a limousine. I shook hands with Charlie Gibson, the moderator for the interview, and on the way to the set, walked past Julia Child (my mother and I watched her on PBS regularly), cleaning up chopped tomatoes from an earlier cooking segment. The whole set smelled vaguely of sautéed onions.

We sat down, shook hands around the sofas, and clipped our microphones on. They asked us one by one to look into a camera so they could take a still shot of everyone's face for their introduction. Then we collectively interviewed the surgeon general on teen health issues. I blushed when halfway through the roundtable Dr. Koop complimented me on a comment I made only because I'd heard my sister make it a couple of days earlier. Then, we all went out to lunch at Rockefeller Center, to FAO Schwarz for photos and browsing, and then to the airport to fly home. I don't remember looking in the mirror once—not even at the studio. The local paper ran a feature, and everyone planned to watch when it aired that Friday morning.

On Friday, I sat in the back of our living room filled with family and neighbors, all waiting for my national TV debut. My parents said they told all of our friends to watch and had lured half of them to our house with a continental breakfast. The segment finally started and each teen was introduced as their headshot flashed on the screen. Everyone held his or her breath in anticipation of mine. Then, there I was, eyes half closed, awkward smile and frizzy newly permed hair extending straight out from my head, beyond the edges of the screen.

My jaw dropped, some of the neighborhood kids snickered, and I thought I saw some of the adults glance at each other. I sat silently on the stairs throughout the rest of the segment and wracked my brain wondering how I could be so unobservant as not to notice that my hair was bigger than Phyllis Diller's, even for 1986. Better yet, why hadn't anyone at the studio pointed it out? Do they think that's just how we wear our hair out here in the Midwest? I had to watch the videotape again later to pay attention to what I said.

I'm sad to say that even now my hair is the first thing I think of regarding that experience. Exciting? Of course. Career-forming? Sure. Life-changing? Absolutely. But the overarching theme to every recollection about it is how I felt when the neighborhood kids giggled at my enormous hair. To this day, everyone who sees the tape says it isn't that bad. Maybe it's not to them, but to me it's absolutely horrific.

Fast forward to a month into my chemotherapy treatments. . . . Seven years later, with all of high school and college behind me, I'd adopted a more mature low-maintenance hairstyle—a long one-length style that went into a ponytail in one swoop. Rarely did I ever primp, and hairspray was a bottle of stinky stuff saved for special occasions. Out of necessity and maturity, much of my vanity and insecurity had evaporated into nostalgia. This is what I thought.

For weeks after my first chemo treatment, my hair screamed at me. It actually hurt. All the follicles were desperately trying to free themselves from the poison nipping at their heels, and I was doing everything I could to keep them in. One morning I stood in front of the bathroom mirror combing my hair after a shower, my scalp aching, and tears streaming down my face. For at least a week, as I lost a strand or two here and there, I ignored the fact that I would lose it at all, telling myself that my hair was stronger and more committed than other people's.

My instinct was to try to glue the hairs back in. So I held the clumps up to the bare patches of scalp hoping maybe they'd grab on again. After a prolonged effort at the impossible, I finally scooped them into my Elvis bathroom trash can—a large tin given to me as a gag gift that once held gallons of popcorn and displayed dazzling pictures of The King in cape and massive white bellbottoms. "He Lives" was painted in huge gold lettering around the top rim, and as I sat on the cold floor staring at it, I knew what I had to do.

Joanie shaved my hair with her boyfriend's beard clippers, as I sat in a chair like an island in the middle of her living room, surrounded by newspapers. This wasn't the first time I'd called Joanie on a whim for an impromptu visit. Sometimes it was so much easier to pass the time driving to and from somewhere, and I knew she'd be able to provide the comfort and distraction I needed, even if for only a few hours. Plus, I think she felt a little honored to be the one to shave my head for me.

Afterward, sitting on the floor in front of a full-length mirror, the baldness seemed almost novel. It made my eyes look larger, sort of like Sinead O'Connor's, which wasn't so bad. My head wasn't shaped like an amoeba and I didn't have any enormous Gorbachev-esque birthmarks, which was good to know. I touched the thin layer of stubble that remained, and it was soft and even like a wellworn teddy bear. When my hand stopped rubbing and touching my head, and I removed it, my head was suddenly very cold.

I reasoned out loud to Joanie, and to myself, that I would experiment with different-colored wigs and buy a bunch of funky hats. I always wanted an excuse to buy a hat rack, I said, and

I've always looked good in hats. I'd even brought a black fedora for the trip home. I said a couple of times that there really was no reason for anyone to know it wasn't a style choice instead of a side effect. The forced optimism didn't stop there. I said that one couldn't see that I wasn't healthy and normal, even though I didn't have any hair and was losing even my eyebrows at an alarming rate. And on and on. After two hours of babbling and reasoning and sitting on the floor in front of that mirror staring at myself, I finally walked out the door into the world.

That weekend I ventured out only once, to the grocery store. I'd already bought my wig at a cancer-patient-wig specialty store in Kentucky. My employer still insisted on skirt suits and pantyhose every day, and I knew that my new classy black fedora would only call attention to my hairlessness. So, I'd made an appointment, sat in a beauty-salon-style chair in front of an enormous mirror surrounded by photos of previous customers who looked remarkably hair-ful—even if most of them wore wide elastic headbands to make sure the wig stayed on over their shiny clean scalps. With the help of a consultant—also a cancer survivor who had worn a wig during treatments—I picked out the style and color for me, an auburn bob.

That Monday would be my new hair's scheduled debut. I got up early and primped and sprayed and moved it around until it swung, swayed, and tucked behind my ear as naturally as possible without the giveaway headband.

Once I got to work I felt better. Most knew of my treatments, but none had seen me without hair, so the exposure was minimal. Several people complimented the wig—what else are they going to do?—but no one stared or snickered. It was all good until, at the exact moment I realized how amazingly itchy my new wig was, I also realized I'd forgotten I had a small part in a new business presentation later that week. That level of exposure was a bit much for my first week out of the closet, but I held back the panic, reminding myself that normal and healthy was all they could see if I had the wig on.

I don't think I remember one thing I said during the presentation. Afterward I heard I was poised and confident, and I even rescued a colleague from certain demise during her portion. All I know is that the president of the company to which we were pitching stared at me throughout. She looked at me, a junior-level member of the team, at least twenty times in the half hour we presented. I ignored her, smiling occasionally, but sweating through the silk blouse under my suit and trying to ignore the itching. I knew I was outed. As we left, we all shook her hand, and she squeezed mine firmly and held it longer than normal. I looked her straight in the eye, but it didn't make any difference. She knew, and my wig must have looked horrible. We didn't get the business.

I hardly ever wore the wig outside of work. I guess I figured that the more normal I tried to look, the less natural I actually looked. The more vain my behavior was, the more it got in the way of my efforts to appear healthy. People looked at me and my head all the time, but the more I worried about it, the more I obsessed about whether people were looking at my fake hair or lack of hair or not at my hair at all, and whether they were pitying me or merely expressing human concern. It had been a long time since *Good Morning, America*, but obviously my hair stories hadn't changed much. What a waste of energy.

I'd wished for no hair, and I was lucky enough not to have to deal with it much for a couple of years. Getting ready in the morning was a breeze. Summer shaving slowed down to once a month, and I didn't even think about eyebrow plucking. I was nearly bald when I first met my husband, and we still laugh because I was wearing a floral dress at a picnic, and he thought I'd been forced to wear it because I was clearly "alternative."

Then it all grew back wonderfully curly and soft like baby hair. Then it straightened and thickened and became exactly as it was before it went away—dark, hard to control and everywhere. But now I don't mind as much. These are the hair stories I want to remember.

chapter five

we've all been there

Working while going through cancer therapy is doable. The comfort of the situation depends, of course, on what kind of cancer you have, what kind of therapy you're completing, what kind of job it is, what kind of people you work with, and, most importantly, how strong your body and mind are. Those are a lot of variables. When I decided to continue working while going through chemotherapy, I thought I had them all covered—a relatively manageable and recoverable chemotherapy regimen, a public relations job that worked my brain much harder than any other part of me, an employer and set of colleagues who were initially flexible and supportive, and a young person's fortitude that wouldn't let some nasty little cancer get in the way of my career or any other kind of goals.

While each variable faltered to some degree during those months, the fortitude one, the one I counted on most, was the one that stumped me.

What I didn't expect was how much my professional life would change. I know, it's hard to believe anyone would expect chemotherapy treatments, not to mention a fight for one's life, wouldn't change things much. But my prognosis was very good and all the work-related variables fell into place so easily in the beginning. I believed that work would be a cocoon where I could maintain that façade of normalcy I desperately needed. Plus, it was the perfect opportunity to show my colleagues, myself, and my cancer that it couldn't stop me. I wouldn't let it take me over.

When I started my job, I was fresh out of college, living in a new city, and I didn't know anyone. People at work buddied up to me a little, but they were still only colleagues I hadn't known very long and with whom I hadn't shared much. Six months later, after working hard and long and barely having time to get to know the city, I was diagnosed and started treatments.

All of my friends lived far away. Some, like Nick, lived very far away. Out of necessity, I grew to depend on my colleagues for support and companionship—at least as much as they could give someone they barely knew.

Mostly, my work companions helped me forget I was sick. They sought out day-to-day conversations, asked me if I needed lunch, criticized my work, disagreed with me in meetings, got angry if I was hogging the fax machine, were glad to see me if I wasn't annoying them, or not so glad to see me if I was. These are the ways that working colleagues think of and treat other working colleagues. It was great most of the time, a true refuge, and everyone agreed that my work hadn't suffered. I think many of them—my office being made up primarily of people between twenty-two and thirty, and therefore practically unable to comprehend a young person's sickness or mortality—were just surprised I made it to work at all.

Jill, my unofficial supervisor at the time, was a single woman like myself who was a couple years older and lived across the street from me. She always was the first to invite me for a drink on Friday nights. She occasionally called on Sundays to ask if I wanted to go to the movies with her. After a while, Sunday lunch and/or movie outings became routine for us. She laughed at my stories about the eccentric pot-smoking cabinetmaker downstairs neighbor and walked quickly up the stairs before he could poke his head out his door. Another colleague, Leeza, and her husband began inviting me for dinner on the Saturday nights they stayed in. Leeza was a marginal cook, but she loved ice skating and ice dancing competitions, so every time one was televised, they had me over. I looked forward to each competition, if only because it got me out of the house.

My colleagues were good people, stymied by the situation of feeling responsibility and sympathy for a woman they barely knew. It always was a bit forced on both our parts—as if we felt we should be good friends because that's what would make sense considering the circumstances. That's what I needed, and they compassionately tried to catch up to meet the need. Every gesture was appreciated, and I was extremely grateful for the companionship. I felt awkward, but extremely grateful. And because of my kind and affable colleagues, I didn't dread going to work.

A few months into treatments, things started to change. Rolling out of bed in the morning became harder work than the job. I grew to loathe the itchy wig I wore only to the office. I was handling day-to-day activities pretty well, thriving on the fact that work kept my mind off what my body was up to, but the amount of work I could take on was falling off. I frequently stayed home on Mondays and left early on the Fridays I wasn't getting treatments. My moods were erratic to say the least, especially the week immediately following each round of treatment.

One day I took it out on Leif, a colleague whose large ego and patronizing manner frequently caused me to ignore his general kindness. I took him into a conference room and noisily berated him for treating me poorly by making a harmless critical comment behind my back. Then I broke down crying and spilled my guts to him for twenty minutes. When my tirade was over, he gave me a hug and left the "meeting" confused about whether he'd done anything wrong.

Another day I wrote a harsh letter to a colleague running a charitable giving campaign in the office. I told him that the force and frequency with which he was asking me to contribute a portion of my abysmally low salary to his cause was bordering on harassment. Since all of my money was currently going to pay the medical bills—the percentage my insurance didn't cover—now arriving daily, I couldn't afford to give one percent, let alone ten, I told him. I would be happy to volunteer some of my time to his organization, but if he didn't stop asking me for money I was going to have to start avoiding him. He awkwardly apologized, provided me with a list of volunteer opportunities, and didn't talk to me much after that.

Another day I was working away, making a string of phone calls, and realized I had to do laundry that night. I had no change for the laundry, and I was physically exhausted. I don't know why, but I turned and complained emphatically to Leeza sitting across the way, who naturally felt obligated to offer her washer and dryer to me for the evening, not expecting I'd accept. I stayed for dinner and until the last sock was dry—11:45 P.M. on a work night.

Some days I didn't even have the energy for emotional outbursts. Some days I was just damned tired, so I got up and went home.

This is neither behavior I'm proud of, nor something that I expected people to just ignore because I was sick. At first they almost seemed to, surprisingly. But after a while, the impositions, quirks, and physical and emotional exhaustion took its toll on my work relationships. It turned out I couldn't hold up my end of the deal. They treated me like a normal, healthy colleague as long as

I remained one, or at least behaved like one. They'd given me the benefit of the doubt for a long while. But I pushed it.

It's true they were my friends, forced or not, but they were my colleagues first. A relationship developed around work will suffer when the work suffers. My work was suffering. My façade had been lowered. My mask of invincibility and fortitude came off. My island of normalcy was sinking and I'd lost control. Maybe I never had any of those things, but at least I could pretend up to then—convince everyone that I could hack it and didn't need much special consideration. Then my days absent increased, and my days leaving early or coming in late increased a lot. I put in the hours each week, but reluctantly.

After a while, the Sunday outings slowed to a halt, the dinner invitations were fewer and farther between. But worst of all, it seemed people stopped trusting me with their work. It started when account teams began including me only as a peripheral member for select writing projects, not integral campaign projects. Then it progressed. I continued to complete my projects, but started needing some leeway on deadlines. Then I started refusing projects altogether because I feared I wouldn't be able to complete them to the standard of quality or timeliness I'd set for myself. Because I was one of two entry-level employees, most of the work began going to the competition, and I couldn't blame them.

As my body continued to feel more and more sick from the treatments, work became a walled-in structure that forced me to pretend I was well. I began recalling more often that the job was only supposed to be a typical first job—temporary, a doorway into the professional world, and perhaps a bridge to a career in writing or teaching. These were all things I was supposed to be figuring out, and instead I was trying to figure out how to survive. My previously pleasant and temporary place of employment became decidedly unpleasant for me, and I was convinced it was showing.

Then one day Jill introduced me to a client who was in the office for a meeting. She was with a marketing research firm for whom I was developing a big writing project when I had to go into the hospital. I completely missed the deadline and, back in those first few days, it was smart of Jill to explain the circumstances to her. I didn't know how much she knew—whether the "c-word" had ever come up—but I clearly was wearing a wig so it wouldn't have been difficult for her to figure out. When we each realized what we knew about each other, we both hastily opened our mouths. I wanted to apologize for missing her deadline and failing her company by not coming through with the document. I started to lower my head in shame and disappointment before I spoke, but the client's words came out first.

"I've thought about you so often since that week," she said, and touched my arm. "How are you doing? You look great, by the way."

My mouth was still open, but I was so surprised that I closed it to give me a moment to figure out what to say. Jill beat me to the punch.

"She is doing great," Jill said. "You wouldn't know what she's been through at all. She's a trouper. We're all very proud of her." I looked at Jill in surprise and waited for a characteristic wink to give away her acting. But she only looked back at me with a little brim of a tear in her eye—yes, a tear.

I stood up straight for the first time in weeks and smiled at them both. I said I was feeling pretty good that day and thanked the client for her concern, and especially for the "looking great" compliment. Then I bounced my auburn bob wig with one hand. They both laughed. I meant every word.

Maybe Leeza with the washer and dryer only remembered that I used to wear a red baseball cap whenever I came over. And Leif may remember that we didn't get along well, but probably doesn't remember the reactionary bitch-and-cry incident at all. We all may remember that we all did pretty good work while we worked together, considering everything. After all, during those years, my colleagues and I got through several weddings, just as many separations and divorces, office romances, client romances, difficult pregnancies, and numerous illnesses. The CEO and founder of the company watched his wife die painfully from breast cancer.

We all had our own traumas, and I don't remember anyone letting anyone else down. I didn't want to be the first.

chapter six

full support without pantyhose

It wasn't a great day, the one on which
I realized I might need some help. The outpatient cancer care unit
was buzzing with activity so it was easier to avoid my nurse than
I'd expected when I started wandering the halls. She undoubtedly
was skulking around, holding up the sterilized L-shaped syringe prepared specially for me.

Every three weeks it was the same. My mother, who usually
accompanied me to the treatments, the nurses, and I all pretended
it wouldn't bother me when she took the funny little needle and
pushed it through the thin skin in my chest—"Little stick"—into the
port-o-catheter sitting above my heart. Then she'd pump in three
bagfuls of liquid of various colors that made me feel dizzy and my
scalp ache. I wouldn't complain at all and would watch soap
operas and read magazines and jiggle my foot for three hours. Then
we'd all smile at each other when I walked out the door because
we'd oh-so-happily "made it through another round of chemo and
everything looks great!" The curse of a good prognosis.

Not only was the unrelentingly positive, cheery, let's-all-think-good-thoughts bullshit making me dream of repeatedly poking my eye with a tongue depressor, but I was having trouble getting motivated to go to chemotherapy at all. So, boredom and the need to avoid the tongue depressor dispenser had me wandering around forcing them to hunt me down.

There was a rack of tri-fold pamphlets littering the front desk. On the bottom shelf was a brochure that seemed to be speaking directly to me (insert face of smoky authoritative *Wizard of Oz*–like apparition here). It talked about hopeful realism as the attitude of choice for "survivors" who refuse to be seen as victims. It told me that informed, empowered, and active was the best kind of cancer patient to be. It spoke of savoring and improving the quality of your life, not just crossing your fingers and hoping your doctors make you live to be a hundred. It offered places and times to talk about what it's really like to have cancer, no matter what your prognosis— no smiling or rose-colored glasses required. I looked up the times and, though I'd never been to therapy of any kind, decided to go.

The center required a preliminary interview, a psychological evaluation, with a staff member. Despite this unfortunate event— "Would you say the cancer has added 'no,' 'a little,' 'moderate,' or 'a large amount' of stress to your life?"—I joined a support group. The program director told me the group had just "lost" a couple of members. I gulped. They were happy to welcome some new blood, especially one so young with such a good prognosis. When I asked if the group was jinxed, the psychologist paused cautiously before chuckling and saying breathily, "No, of course not. . . ." I didn't take it as a good sign.

I was the first to arrive since I'd left home

about a half hour too early. Because I'd never been to therapy before and knew that I probably should have made it part of my personal development long before this day, I feared that if I weren't prepared and alert, twenty-two years worth of "issues" would come pouring out. So, I spent the drive rehearsing my glowing yet humble personal introduction, meant to assure the group that I was confident of my excellent prognosis and that I was really only there to have somewhere to go on Tuesday evenings where I didn't have to wear a wig. Of course, I added that I was there also to "help others not doing as well as I am." How charitable of me.

I was running through it again when a gray-haired stout lady with a bandage on her neck walked in. She walked up and thrust her hand in front of me, opened her mouth and scratched out some version of her name I couldn't understand, and sat down right beside me. A dozen other softly lit chairs, including very plush recliners, surrounded the room, yet this woman chose to sit only inches away from me. She was pulling out about a dozen cough drops and a box of tissues. Did I look like I was going to bust like Hoover Dam?

One by one the group came in—most at least twenty years older than I and only one who actually looked like he had cancer. He had black baggy circles under his eyes, hunched shoulders, and a colostomy bag he carried around like a trophy, thrusting it forward when he told the man sitting next to him the latest bad news the doctor had given him. The group was quietly familiar in their greetings to each other, and we sat nearly silently until our facilitator entered.

Vic was a broad man who looked like he was most comfortable in a sweater and jeans, but had resigned himself to wearing a tie almost every day of his life because of his professional choices. He had practically no hair, and where he didn't there was a well-scrubbed, well-oiled scalp to dazzle the beholder into ambivalence about the lack of coverage. His face was peaceful and pleasant, and he made the whole room smile when he walked in.

Since there were two new members of the group, Vic prompt-ed us to go around the room and introduce ourselves. The stan-dard information given in lieu of rank and serial number was: Name, Age, Diagnosis, Date of Diagnosis, Treatment, Prognosis (if known), and Length of Time in the Support Group. Nearly everyone gave nearly all of this information in that order. The pros listed it all like it had been tattooed on their arms and the only thing they had to do with their time was memorize and practice. I didn't want to, but I found myself wondering what else there was to everyone's story, especially when they offered a taste—like the "fifty-one-year-old lumpectomy divorced former nun with breast cancer." I figured there had to be a great story there.

It came around to me, and I gave my little rehearsed speech and everyone welcomed me. Then Vic asked if anyone had a spe-cial need for the session, anything they wanted to talk about. I realized then that Vic didn't do much of the talking, the group did, and those were the folks I should be wary of.

A very quiet voice came from the sofa in the corner, and the divorced nun spoke about how her last visit to the doctor had revealed her post-lumpectomy radiation treatment wasn't working, and she'd have to go on Tamoxifen, possibly for the rest of her life. She didn't cry but she was obviously upset so everyone offered her supportive words. Vic helped her talk through her fears as long as she needed; then we gladly changed the subject. Next was Joey, the "forty-six-year-old colon cancer patient lives with father colostomy bag wearer." He almost smiled when he told us that he had been feeling very weak and his doctor was probably going to put him into the hospital to rest and get some fluids. Finally, in what can only be described as classic passive aggres-sion, he refused anyone's sympathy.

Then Gert, the "eighty-one-year-old smoker lung-and-throat cancer" patient sitting ever so close to me spoke up—as much as a person without a chunk of her larynx can. Mostly, she just sat and cried and babbled hoarsely, and we all pretended to understand. Most of the room reacted in mere repulsion from her incoherent display, no one reached out to her, to touch her hand, to offer a hug, nothing. She was the only one to cry, and I now knew why she came prepared with a whole box of tissues. But even the others in the Bad News Crew were comforted appropriately, not gushingly but thoughtfully. For Gert, nothing. Then, through the silence came a vibrant voice at the other end of the room.

"Well, for god sakes, I have some good news if anyone wants to hear it. I know I could use some," said the "forty-one-year-old double-mastectomy breast cancer survivor," with no recurring symptoms and excellent prognosis. She was thin, nicely dressed, with short dirty-blond hair and a constant full-on smirk, meant to make you think she knew something no one else did. In our introductions she'd described herself as "interested in the human condition," seemingly trying to prove that as her motivation for attending weekly for more than a year, even though she'd been cancer free for most of that time. It was clear she wasn't giving up anything and had no intention of wallowing in bad news. I was immediately drawn to her.

Everyone turned their hopeful eyes her way, and she proceeded to tell the most fantastical story I'd ever heard from a cancer survivor, offered as nonchalantly as if she were ordering at a diner.

"My boyfriend Rolland, you remember the guy I told you about, probably an alcoholic, well he finally left . . . in a patrol car . . . after he pulled a gun on me and Chelsea, my seventeen-year-old daughter for you new ones. He was sober for a while after I told him to go away or stop drinking, but then he started again one night and told me I was too goddamn lazy to even give him a blow job and he was angry because we hadn't had sex since they cut my boobs off. I again told him to leave so he pulled out a gun, swung it around a bit, I tried to talk him down, all the time wondering how he'd hidden a gun from me for two years, but finally the neighbors called the police because of the yelling and they busted in and took him away. I'm sure we'll never see him again, and we're finally rid of him. . . . Isn't that good news?"

I looked around and was dumbfounded to find that I was the only one who was awestruck and completely fascinated. The group immediately and casually congratulated her, and Vic seemed the most pleased.

"That's so good to hear, since the last we heard you were thinking about asking him to leave," Vic said. "Did our talking about it spur that on? How do you feel?"

The woman gave Vic an eagle-eyed look that should have made his balls shrink about six sizes and said, "How the hell do you think I feel? I just had my asshole boyfriend arrested, and I cheated the devil yet again out of having me at his hot little party in the basement." Everyone, including Vic, just smiled and chuckled softly. I'd become part of the party and knew exactly who the life of it was.

My third week in the support group, there was a reception in the lobby afterwards for the groups meeting on Tuesdays. On my way to the punch bowl, the fascinating breast cancer survivor, now known as Ginnie, short for Virginia, came up to me and said, "I can't stand it here another minute. Let's go." We walked to the restaurant next door. Without skipping a beat she ordered a drink and then told me all the dirt on Vic and his messy divorce. She rolled her eyes when I asked about Joey and told me how he constantly says he's hanging on to life by a string but manages to attend every week without fail. "I think he hangs on just so people will feel sorry for him," she said. We talked for quite a while about Gert and how every week she tries to dominate the conversation with her pity party and how often she always "needs everyone's support more than ever." Ginnie told me that Gert smoked two packs a day for sixty-five years, still smokes every now and then, and has run everyone out of her life with her general nastiness.

I resisted the urge at first to ask Ginnie what her own "real story" was. She hadn't asked mine so I thought it overstepping my bounds. But an hour later she surprised me. She looked over the celery stalk in her Bloody Mary, gave me a different version of the eagle eye she threw at Vic, and asked me directly why I was in the group. I was a mystery to her too, apparently. I loved it.

"You never say anything, except to tell everyone how freakin' well you're feeling," Ginnie said skeptically through her eyebrows. "I see those dark circles under your eyes. And tonight you welled up when Ashley talked about how she dreamt last night that she had never gotten cancer." I couldn't look her in the eyes, so I stared at my finger stirring the foam in my beer. "It's killing you being here every week, isn't it?" Stirring. "Why are you here?"

I had no answer for her. I had a system of friends and family to depend on, especially Sean who was being generally great, considering I probably asked him for too much, so it wasn't that I needed more people to be able to call late at night. Maybe I just wanted to find someone who would be able to nod knowingly when I described a feeling that seemed too extreme or a reaction that appeared too ungrateful for someone who'd never gone through anything like it. Everyone outside the group was so focused on what was going well for me and how grateful they were "this cancer thing" wasn't as big a deal as it could have been—caught early, good prognosis, "easy" treatments—it was nice to be somewhere where they almost expected me to be doing badly, where they let me be sick.

Despite all this, before that night I had raced out of the room at the end of each session in order to avoid post-group contact.

Over time I began genuinely caring about the group, but not enough to want to go through the torture of attending every week. It was helpful for the others to see someone appearing to be doing so well, but I had to admit I wasn't that altruistic.

But on Tuesday nights I couldn't hide or pocket my issues away neatly. I'd joined the group because I thought I'd be able to be honest about how horribly I felt most of the time and how guilty I felt about being so depressed even though I was probably going to survive. But every week I choked because my reactions were too intense to handle in front of people. Among the group, I was reminded of how different we all were from almost everyone else in our lives. When we were together, we were diverse but normal. When we walked out, the lack of normalcy was glaring. We all became someone's tired, bald friend/relative/neighbor/acquaintance/employee with cancer. Someone who people felt sorry for or worried about or, if you didn't have any hair, stared at.

At a precancer twenty-two, my own skin and body, not to mention everything going on in my head, hadn't become anything approaching comfortable yet. With cancer, it already seemed old and sick and tired. I hated being reminded of that. I hated being reminded of any of it, and part of me missed the rose-colored glasses they passed out liberally at the hospital. Most of the group seemed very comforted by the meetings, and in certain moments I was, as well, but most of the time they only pointed out an inexcusable amount of pain in one room. I couldn't see anyone who was "empowered and active" as promised, especially not me. But I may not have been looking. I may not have been ready to see it.

The next Tuesday I showed up again. At
least I knew I had someone to have a drink and share nachos with following the torture, and she was vastly entertaining. She

defied convention at every turn. She started her own million-dollar data processing company in her twenties, but she'd only graduated from high school. She was married for a year, had a baby, then divorced, and is one of those few astounding people who make us all look bad by achieving a normal, healthy relationship with her ex. She spends every other Christmas with him and his wife, whether or not her daughter joins them.

Though she was skilled at avoiding offering personal information considered too sensitive for my young ears, there wasn't much she wouldn't say. Her mouth simply had no censoring mechanism, and it helped me relax. In my world filled with people who looked at me too much but were careful not to outwardly recognize that I had no hair, she called me LBK—Little Bald Kid. I admiringly called her a bitch when she deserved it, and we generally practiced the "just say it" method of communication.

Every week, when we were done debriefing about the earlier group meeting, she'd always ask me why I didn't bring up [insert my exact problem that day here] during group. Then we'd talk about it for hours, but never about her problem of the day.

One week, Ginnie finally and abruptly started talking about herself. Her "segue" into Ginnie Time was nervous and rehearsed—a quick flick of her hand to the bartender as she swallowed the last of her Bloody Mary, then a slam of the glass on the table and a sudden regurgitation of the gist of the problem. This first night it was her mother—"So, did I ever tell you about my horrible mother?"—then on to tell me about her recent death and her vow to never be anything but kind to her teenage daughter.

I was silent for a long time while she talked. Then we were both silent. It hadn't taken long, and I knew that she was done talking about it. A surprising wave of relief washed over me, not because of the tense moments surrounding her visibly uncomfortable purging, but because I was happy. I'd finally found my cancer support group.

chapter seven

praying on my conscience

If you're in a cancer support group long enough, you'll go to at least one funeral. I went to two in my nine months in the group. Ginnie was in the group for almost two years, and she went to four. It's not like a person in a cancer support group dying is a huge surprise for anyone. But these were the most difficult funerals of my life, especially the first one.

About four months after I joined the group, Ginnie moved away—back, I should say—to Albuquerque where she grew up and had lived most of her life. She'd moved to Ohio so her daughter could be close to her father, but now Chelsea was going to Arizona for college. So, back to the vastness and familiarity of the Southwest she happily drove with everything she owned in the back of her truck. I envied the fresh start and clean bill of health she had. I also lamented that, as my cancer support group, she would be too far for a weekly commute. We would talk regularly, but Tuesday nights' group meetings would again become self-inflicted and guilt-motivated torture that I felt absolutely obligated to endure.

I'm not sure if it was a coincidence that the first Tuesday after Ginnie left was the first meeting J.T. attended. He was sitting in the low-lit meeting room alone, except for Gert who'd planted herself right next to him, when I arrived that night. It was to be my last meeting, I'd decided, though I hadn't yet masterminded the excuse I would give. Now, the only other under-thirty face I'd seen since I joined the group distracted me. He was tall and gaunt, and I couldn't tell if he was really sick or just thin by nature. His brown eyes were younger than his body and his chemo hair was down to light-colored wisps of a receding hairline. He didn't seem phased by the fact that Gert was sitting mere millimeters from him, and he watched her prepare herself for her weekly self-pity purge with a slightly amused smile and not a word.

After everyone's now-familiar recitation at introduction time, J.T. introduced himself: "Hi, I'm J.T. and no I don't want to tell you what it stands for. I'll start with the important stuff. I'm married and have a two-year-old son. I manage a shoe store in the mall and run the youth group at my church with my wife. Born and raised in Ohio, and even if I do beat this thing, I'll probably never leave. Been fighting leukemia, the kind kids get, for about three years now, but I figure if kids get it, it's probably been in there a while." He'd tried chemo and radiation that seem to keep the cancer from growing, but they weren't kicking it either. The insurance company just loved him because he'd just gotten out of the hospital for the sixth time because he was so anemic that the doctor wanted to pump a transfusion—some "lifeblood"—into him before letting him loose on the mall again.

"I've gone to church all my life," he said, as we all sat in silence listening to this story he told with an anxious smile that turned down a little. Then he spoke more slowly. "I'm God-fearing, God-abiding, God-loving. My entire family and church pray for me. My wife and I pray every night until our hands cramp up into fists. But even all that praying isn't enough to convince God to make me well enough to play with my baby son in the park. I'm not dead, but I'm not well either, and I'm very, very tired. Sometimes I'd rather be dead." No one had any idea what to say, so no one said anything. J.T. kept going.

"So, I'm here to see if this is what I'm missing. I need the group, and I need some of that empowerment the brochure talks about. Basically, I need you guys to help make me well, or at least feel that way. And I promise to try to do the same for you."

He didn't want to reveal to us what J.T. stood for, but the newest member of our little group exposed more about himself than any of us had dared in the three months I'd been attending. It was refreshing and wonderful . . . and my stomach dropped out from under me because I knew I'd be devastated if he died. I probably should have been ashamed for so desperately wanting him to live so that I wouldn't feel like a failure, but I was over-powered by this incredible feeling of being needed and wanting to protect him from harm. I immediately adored him.

Ginnie always had reported getting regular phone calls during the week from group members looking for supportive words or advice—usually Joey, looking for someone to feel sorry for him. But my first call didn't happen until J.T. joined. Then a couple of weeks into his membership he began calling regularly—almost every Tuesday, like clockwork, after group. It didn't involve cock-tails and nachos, but it was a version of the Tuesday night outings I'd become accustomed to.

He was hesitant at first, like Ginnie, only talking about the group and their quirks. But he obviously had reasons for calling. He had a good marriage and a happy family life, but, like me, there was something he wasn't getting from the group or anyone else. Suddenly, the group's responsibility was all mine. And J.T.

didn't seem to have much time. I finally just asked him why he called. Over the next few weeks, he answered.

J.T. said I was the only one in the group he identified with because I was younger and hadn't lived most of my life like the others—we had a right to be angry and frustrated with the prime of our lives being interrupted like this. I said maybe this was the "prime of our lives," the time when we discover who we are and what we're made of. We were just lucky to be able to define ourselves earlier in life than most people ever have a chance to. He said he refused to be defined by his cancer. I said I hated that it could have that kind of control over my identity.

He talked about how the group never really talked "about" anything important, just regurgitated the latest news from the doctors and listed new treatments or medicines they're on, and that he couldn't believe a shy Ohio country boy was the most expressive in the group. I said we admired and feared him because he was honest enough to tell us how much he needed us.

We often made fun of Lila, a very kind and proper new member who showed up every week in an outfit that was probably extremely expensive in the late '70s when she bought it, and we'd dissect how much that week's ensemble probably cost. We commented on the Gilligan-esque floppy hat one of the young ones from the other Tuesday night group had worn and thought it would be excellent for grocery store outings. He said he kind of admired Vic's boldness for shaving his remaining hair and showing up sparkling clean bald to a center where most everyone was bald against their will. Healthy people spent a lot of time worrying about their clothing and hairstyle choices, and we envied them.

He never wanted to talk religion, but I did anyway. I told him that I grew up in a religious home, too—church every Sunday, youth group every weekend, a prayer before every meal. Most of my dreams were set in my family church, I said, no matter what the plot. One night that week I'd dreamt a date took me to dinner at a restaurant in the foyer to the sanctuary, complete with sophisticated couples with cellular phones and designer clothing. Sometimes, I remembered people I hadn't thought of in years because I dreamt I ran into them at a Sunday morning service there. They'd never been anywhere close to the church, I was positive.

J.T. said that church was probably comforting for me—a safe refuge. I said I never went to church anymore and would feel like a hypocrite for praying. He said there were a lot of people praying in the world—enough to make up for God not hearing from me much anymore. I knew he was right. Once during college, I'd gone to rural Northern Ohio, to my friend Nick's hometown, for the weekend. His parents were devout Catholics and believed in the healing power of prayer, especially when it came to Nick's mother's crippling multiple sclerosis. As I hung my head over the toilet after Nick poured me too many Seven & Sevens, all I could hear was a low murmur from the next room. Through my drunken haze, I realized it was his parents saying Hail Marys over and over again, overlapping each other, almost chanting. I imagined them kneeling next to the bed, white-knuckled fists clasped together, their faces scrunched up with fervor and heartfelt penitence. I told J.T. that I felt very sorry for them. He didn't say anything.

Mostly, we talked about how we both still worked, how we felt like we were faking the whole thing by pretending it was as important to us as it was to our colleagues, when all we really wanted was to keep our paychecks and health insurance, not feel useless, and not let anyone down. His coworkers were great, J.T. explained, but he knew they hated him for reminding them that young people die too.

We both knew that it wasn't good how tired we felt at the end of the week, especially J.T.

I knew that Ginnie would really like him, I said, and maybe he should take some time off and get some rest.

He said he worked so he could build his savings so his wife and son wouldn't be stuck without anything.

I said he wasn't going to die yet.

He told me that he loved his family more than himself, but he couldn't talk to his wife about a lot of things, especially how desperately he didn't want to leave her to raise their little boy alone. But he was very tired.

He never said he knew he was going to die soon, but he probably did.

It was all very expected and cliché the
way it happened. J.T. probably liked it that his story was perfect for a movie of the week, but I was beside myself with agitation that it all seemed so contrived.

One week J.T. didn't show up to group. The next he left word with Vic that he was in the hospital. I went to see him on one of his good days while his wife was at work, and he said he didn't like it that I saw him in the hospital. He was feeling okay; "realistically optimistic" were the words we learned in group, and I could tell he practiced them when people weren't around, like I did. I was hurt that he used them on me rather than telling me how scared he really was. It was excruciating sitting there watching someone only five years older than me look and feel like an old man, and I understood why no one but my parents ever stayed that long when I was in the hospital.

There was a crucifix on the wall above his bed, and after an awkward lull in conversation, I asked him if he still prayed until his hands were cramped into fists with vehemence. He said he'd left the praying to his family and friends who seemed to feel better for doing something to help. I told him I'd be thinking of him and looked forward to our next Tuesday evening phone call. I couldn't bring myself to tell him I would pray for him.

I tried to call him later that week, but his wife said he was too tired to talk to any of us. I wasn't quite sure who "us" was, but I guessed she was referring to the support group who had probably all called. I found myself pleading with her for just one last word, finding the possibility of not saying good-bye unbearable. I wanted to tell him I'd miss him and that he was a good friend. She just said he was sleeping, but thanked me for calling.

The group sat together at J.T.'s funeral,

a fact that made it impossible for me to consider leaving the group any time soon. Who else could I talk about this with? Who else would understand? We were there an hour early, and we still had to sit in the next-to-last pew—our punishment for having failed him. About fifty pews and five hundred people back, we couldn't even see J.T.'s wife and son. I'd never been to a funeral for a person who died during their prime, leaving everyone behind instead of joining them somewhere.

We sang six hymns, heard three scripture readings, recited four prayers not counting the Lord's Prayer, listened to the children's choir and the youth choir tearfully sing "Friends Are Friends Forever," took communion, and listened to the minister's full-length sermon. I'm not sure if anyone except me knew J.T. had given up praying when the minister called "Jim" a "remarkably devout man," but I'm sure they all felt better for it. At least they knew he was in heaven and they'd "see him there when God decided it was their time to be with him." I wondered how J.T. was explaining all this to God.

I wondered if I'd see J.T. again and if it would be in heaven. He'd stopped praying, but maybe God had heard his prayers and was hanging on to them until he saw him, just to prove it. Maybe J.T. died because his body was too weak to fight off the childhood leukemia it unleashed only after he'd fathered a child. Maybe he is in heaven, and maybe there is no heaven. Maybe I'll see him again, and maybe I won't.

Either way, I know that J.T.'s death was much more than a cliché. I'm sorry I failed him.

chapter eight

the rebellion

My life, my future, is my own, and you can't tell me any differently. In fact, I've spent many years rebelling against anyone or anything who ever tried to impress upon me that my destiny is not in my own hands. My parents say I was born with my fist in the air, declaring my liberty and with a righteous scream to match. They should know.

My parents became accustomed to my independent wanderings when I was a child. The most notorious incident occurred when I was seven, and I wandered off at the crowded Columbus Zoo for a good hour, my parents racing around, grabbing security guards, splitting up to search the entire zoo. They found me in the last place they'd imagined a little girl in double pigtails tied with little yellow bows wearing a yellow and pink polyester shorts outfit would go—the snake house. Honestly, the snakes were fascinating and not scary at all to me, but that wasn't the only reason I went there. When I finally wandered over to my father racing in the front door searching for me, I knew I'd found the one place he wouldn't think of telling me not to go because he thought I'd never have a burning need to go there without supervision. It was a twisted and selfish sense of satisfaction, but satisfaction, nonetheless.

My father should understand the consequences of this independent streak more than most. He bore the wrath of my metaphorical fist and unfortunately authentic righteous scream more times than my mother or sister care to count when I was a teenager. It would take the fingers and toes of several people to count the number of times I stood over my father sitting in his chair, screaming at the top of my lungs in outrage over some rule or curfew I was convinced was absolutely unreasonable. Just as many times, he sat in the living room, me sitting defiantly with my arms and legs crossed in the love seat in the corner, as he told me he was very disappointed because I blatantly broke that rule or curfew. No matter what, I was not to be controlled. Period.

I still have a marked disdain for authority. It's not that I necessarily dislike those who have any sort of control over my life—employers, landlords, police officers, legislators, or half-boyfriends. It's just that their control means that somewhere I lack a measure of the same. That is unacceptable.

It's not just for me that it's unacceptable. The causes I find myself gravitating to are those associated with fighting for the civil liberties of others. In my white middle-class existence I have rarely been oppressed, and never in a grave manner. But I have gladly taken on the fight against oppression of others—marched on Washington for abortion rights, attended and even performed at civil rights rallies, helped set up meetings with residence hall leaders for hallmates who felt the rules of the residence were too strict for their lifestyle. If it's possible that I or someone I love or even someone I have never met will some day be trapped in a Thai prison or require postcoital birth control, I want to know that our rights are in place and well protected by an army of like-minded folks. Anything less is unacceptable.

The moment I was diagnosed, I was forced, kicking and screaming, to hand over control of my immediate future to a strain of mutant cells traveling through my lymph system. As you can imagine, this didn't go over very well. I woke up sweaty at night, imagining them racing through my bloodstream, cackling and dragging the shackles intended for me behind them. The "good" poisons of chemotherapy treatments weren't any better—they just didn't mean it. They couldn't help but debilitate my immune system on their way to the cancer cells, making germs and infections a serious threat to that control. These were entirely unacceptable authority figures. Was I going to let my body, mind and life be lassoed and calf-tied by some hungry little goblins with an appetite for cells? Never.

Suddenly, my battle against the disease wasn't about trying to kill it. It became a hair-pulling, teeth-gnashing, blood-vessel-in-your-neck-popping tug-of-war for control over my own destiny. It didn't even have to be about "life" or "destiny" or "future" all the time. Half the time, it was about small things—an afternoon I didn't want to go to the treatments or I needed to go to the grocery and I couldn't get off the sofa. Sometimes it was about bigger things—giving up my cat or not being able to change jobs because of my health insurance. I didn't have to give up my cat, and who knows if I would've changed jobs at the time if I could have. It was the premise, the fact that my options were slimming daily. The lymphoma and all the creatures that came with it were the embodiment of what I'd rebelled against all my life, and it was nicely packaged in its own ecosystem . . . me.

Finally, I found the chance to gain some control back.

Sean asked me to accompany him on a postgraduation backpacking trip to Europe he'd planned for years. He was going around the time I would finish my first seven—and possibly last seven—rounds of chemotherapy, so he billed it as a celebration of sorts. But I think he really wanted to give me something to look forward to, to force me to set my sights farther than the immediate future. Something I'd been unwilling, and perhaps unable, to do.

My first reaction was, "What a great idea, but there's no way." I'd barely be finished with treatments, and who knows how I'd feel. Then it came to me: a trip to Europe was the ultimate rebellion. It was totally against doctor's orders. It was contrary to logical reasoning. The deciding factor was that I knew the cancer would never expect something so extremely devil-may-care, quality-of-life oriented. It would be like dating the bad boy in town just to rile up your parents. I'd take those nasty cells from the high road, jump on to the horse from above, and have a once-in-a-lifetime experience despite it. It was the perfect ambush. Besides, I couldn't think of a more romantic place to spend time with my half-boyfriend.

About three months before the trip and only a few months into treatments, I started telling people about my plans. I watched one chin drop after another. My parents looked at me blankly, knowing their words would be useless, and then asked what my doctor thought. My doctor quickly picked up his chin and smiled, remembering my reaction the time he "strongly recommended" I stop drinking alcohol altogether, and provided me with a prescription for antibiotics in case any infections should crop up. My support group cheered me on in spirit, understanding the impulses I was acting upon, but then continued with subtle comments about caution and taking care of myself. Sean was happy but surprised, perhaps never really believing I'd have the balls to do it and suddenly not sure how ready he was to go on his carefree, celebratory European backpacking trip with a hairless cancer patient. The guy who took my photo for my new passport accidentally touched my wig as he unrolled the backdrop, looked at my pale drawn face skeptically and then asked me how long I'd be overseas. Two and a half weeks, I told him, long enough to get into a lot of trouble. It also was long enough for me to get really sick.

The trip approached. My chemotherapy was going well—the tumors continued to shrink, and my blood counts indicated that the lymphoma was in remission—but I was feeling like I'd finished an Iron Man competition. The few cancer cells that remained were fighting the lingering poison for some measure of control, and I was fighting them both for it all. My body ached, even at the end of a very easy day. I required a lot of sleep, and I had lost some weight. Technically, I was winning, but I'm not sure by what margin. It didn't matter. I was determined to make and enjoy the trip, specially planned to contain lots of European wow-factor. Paris and Notre Dame, Nice and rocky beaches, Rome and the Vatican, Florence and Michelangelo's David, Venice and watery charm, Munich and beer gardens. No type of cell was going to keep me from them.

It wasn't long until something went wrong. Okay, it was only our first day in Paris. We got to our hotel around eight o'clock in the morning after an all-night flight. We were staying in the trendy Latin Quarter, and from the balcony we could see the Eiffel Tower in the distance to our left and Notre Dame just across the Seine to the right. It was beautiful and magical and so gosh-darn European.

But I was exhausted, and we both stank. I was going to take a quick nap and have a shower before heading out, I said, and went to bed, as Sean went out to explore. Twelve hours later I woke with a body ache I'd spent months fearing and watching for. I knew I had a fever. I sneaked to my backpack for the antibiotics my doctor gave me, shakily opened the bottle and swallowed two with a dry mouth just to get them in there as quickly as possible. I lay back down and thought of things I could do to help—baths, cool towels. I remembered a workshop I'd attended on imagery and meditation and moved my arms to my side and tried to relax and imagine the antibiotics hunting down the stray bacteria and attacking them at the jugular and shaking them until they didn't move. I found these images decidedly not conducive to relaxation.

Instead of relaxing and imagining warm sunshine and happy places, I wanted to bang my head against the wall or throw the pills off the balcony onto the charming Europeans strolling to dinner below. I thought about hopping a cab back to the airport and leaving Sean with a resigned and pathetic note on his pillow: "I'm sorry, but I'm leaving you alone in a strange country because I was too proud and brainless to realize that the fact that I have cancer and have been systematically poisoned for the past five months may mean I'm not strong enough to withstand budget-constrained transcontinental travel. See you back in Ohio. Pathetically yours, Stupid." A part of me thought he might enjoy his trip more without having to drag my tired, sick ass around behind him. I wouldn't have been surprised if he'd had the same thought at least once already.

I stared at the dim Parisian lights, a mixture of pinpricks and dull aches pinning me to the hard mattress and thin pillows. What was I thinking? Was I truly trying to "live life to the fullest" or was I going to kill myself out of pathetic desperation to be the master of whatever life I had left? What was Sean doing out there without me, and why was I so dim that I thought I wouldn't mind if he was out there without me?

That snake house rebellious spirit was still in me, but I almost wished it wasn't. I was appalled at the discovery that going to Europe in a body that's been an unwitting battleground was the length I was willing to go to self-righteously thumb my nose at my disease. And it wasn't just the cancer, but my doctors, my family, Sean, my concerned friends, all the people whose chins I picked up with glee. They also wanted to control the cancer inside me, but, unfortunately, it came in a package with a willfulness that had its own agenda.

I should have learned my lesson about underestimating the risks on that lake in Connecticut with Nick, just over a year before. Instead, I was laughing yet again in the face of death just to prove it couldn't get the best of me, and look where my rebellion had gotten me: sick in Paris with a beautiful view that I couldn't see lying down, missing a half-boyfriend who's out seeing the city and meeting beautiful women at my insistence.

Twelve hours of restless sleep and cool showers later, the fever went away. I took the antibiotics and rested in that lovely room, smiling pathetically at Sean when he returned. I started slowly on my journey to full-fledged tourist with a short trip out for a baguette and some tea at a café around the corner. Then I moved on to sitting bundled up on a bench in a beautiful little park across the Seine from Notre Dame, writing in my journal and taking photos of children playing. Eventually, I stood in front of the Mona Lisa, climbed to the top of Notre Dame with Sean puffing behind me, and watched a thunderstorm from the top of the Eiffel Tower. It was my favorite leg of the trip.

The rest of our tour was great too, but Paris was where I stared cancer cells in the face and conceded that they might have a point. Paris is where I discovered that the cells weren't stronger than I, but that they most definitely commanded—no, earned—my respect, as any good authority figure does. I'd go back to Paris in a second.

It's good to struggle against authority—to stake your own ground and have something to succeed in spite of and, in some cases, because of. It's even good to be angry at it at times. It feeds the fight. It helps you win. But never underestimate it. After Paris, I never did again.

chapter nine

french kissing

France is a strange and beautiful country.
It's romantic, but in an oddly practical way. Romance is just what
they do and how they are. They expect romance and to fall in
love or desire to fall in love or to make lots of love. It's just a part
of their everyday culture, and they are comfortable with it.

That's swell if you're young, healthy, and beautiful. When I
was there, I had one out of three going for me. I'd just finished
six months of chemotherapy (~~healthy~~), so I was pale and bald
(~~beautiful~~). I did have "young" going for me, but when you're basically unhealthy and feeling ugly in one of the most romantic coun-
tries in the world, being nearly twenty-three just doesn't matter.

The fact was I'd given up on sex being a big part of my life
the day I was diagnosed. Sure, Sean was still around, but the few-
and-far-between "comfort sex" we'd had initially came grinding to
a halt as the treatments progressed and our relationship became
more complicated. It was becoming difficult enough for us to be
friends, let alone lovers, Sean said, after rejecting me for the first

time. He was right. But he was my last chance for any kind of relationship throughout all of this. I wasn't dating anyone else at all, and couldn't imagine it. After months of chemotherapy, I was too tired to go through the rigmarole of a new relationship, too exhausted to be interested in anything remotely sexual, especially with someone new. And frankly, I would have been surprised if anyone were interested in me, a woman who not only had cancer but also looked like she had cancer.

But I was young. I was tired, but my hormones weren't slowed by some pesky poison or mutant cells. I missed it—not just the sex, but also the intimacy. Most of all I missed kissing. I missed everything that a good, warm kiss can communicate—the subtle, and sometimes not-so-subtle, sexuality of it. That rush the moment you feel someone is interested in more than your mind.

My first real kiss had been a very innocent one given only after months of hand-holding and arm-over-the-shoulder walking with my twelve-year-old boyfriend. It happened on my back porch during the spring of my seventh grade year, the boy I was "going with" still wearing his paper route bag, and me in my favorite jeans and sweater with red and white stripes (like a peppermint stick) worn for the occasion. We'd been sitting holding hands and being nervous about being alone together for at least an hour by then, and he had to get home to do his homework. So, we stood up and he put his arms around my neck, raised up on his toes a little—he was the same age and therefore shorter than I—and came toward me with closed eyes. He didn't check if my eyes were closed yet, or even if I wanted to kiss him; he just assumed I did since I'd willingly held his hand for so long. So, I closed my eyes quickly to prepare and suddenly our lips touched. His were soft and thin, a little too pursed from nerves and a little tickly from the pubescent mustache growing on his upper lip that he talked about a lot but I never saw before I was within kissing distance. We weren't sure what to do once we were there, so we did nothing—just held our positions until one of us decided it was time to pull away a few seconds later.

Not one to put on the Most Exciting Kisses of All Times list, but very sweet nonetheless. The best part was when we pulled away. His eyes, down-turned gray ones with dark eyelashes, were only an inch or so away from mine. I'd never been that close to anyone's face before and I found it very intimate and, in my innocent twelve-year-old mind, sexy. He must have too, because he turned away from me quickly, pulled his newspaper bag down so it covered more of his front than his back, and barely said good-bye before racing away down the driveway. My face flushed, and I didn't stop smiling until hours later.

That's the kind of simple, intimate rush I missed. I was sad I'd have to wait until I was healthy and desirable again to experience it. But for some stupid reason I went to France anyway.

The beach in Nice, France, is rocky, hot,

and scattered with as many kinds of people as the world—old ladies, very young men, groups of young beautiful European women, families, newlyweds, middle-aged couples, French, South American, Australian, Italian.

It's amazing anyone goes there at all because you have to buy or rent a heavy bamboo mat to sit because the rocks are so large and hot that it's impossible to be comfortable otherwise. Vendors walk around selling beer and bread, and hardly anyone actually goes into the ocean. Believe me, it's much more pleasant to look at it than to touch the turbulent water and rocky, unpredictable footholds. Just sitting on the small rocks on the edge of the surf our first day there, the crotch of my bathing suit filled up with pebbles and bagged like it had a two-pound weight strapped to it after only five minutes. I decided not to reach down and clean the rocks out of my ying-yang in front of hundreds of people. I had to get Sean to bring me a towel to wrap around my waist so the entire beach didn't see all of me as I returned to my mat, and then had to get him to stop teasing me about it for the rest of the trip.

Oh yeah, and by the way, on the French Riviera, most of the women are topless. It wasn't just the young nubile women happily flinging their bikini tops off. It didn't matter how old, young, small, large, perky, or saggy they were. It's what you do on European beaches—without big notices and fences to protect children's eyes. I have evidence. I took a picture of the beautiful rocky hills that jut out from the shore down the beach and inadvertently captured the full front of an elderly woman who apparently had many children, all of whom were breast-fed until they were five.

So, I know you want to know. What'd I do under the social pressure of the French Riviera? Let's just say I was glad to have the port-o-catheter still protruding from my chest as an excuse not to bare my own smallish numbers. After wearing tank tops in the hot August weather for three days already, I knew that few people would even notice the bulge, and wouldn't care if they did. The French were sensibly aloof—not snotty or mean as their reputation suggests, just decidedly within themselves when by themselves. I appreciated that. Nevertheless, I did not want to take my top off.

Don't worry, I didn't disappoint. On my body remained a one-piece blue bathing suit, but on my head was only a coordinated blue cotton bandanna. For months, the only people who had seen me without a bandana or hat on my head were Sean, Joanie, and my family. This was by choice. It was not more comfortable with something on my head, except at night when my head got cold, and anyone with eyes could see that I didn't have any hair under my adornment of choice. It was just too much, too noticeable, and too outwardly sickly to go out in public bald.

I hadn't gone to the beach with the intention of taking my bandana off, but looking at all the women I began to lose my inhibitions. I hadn't felt fresh air on my scalp in months, and I thought it appropriate that it was the same salty air that touched the variety of bare breasts surrounding me. It took a few tries, but I finally stripped the bandana off and sat, bald, looking at the dark blue ocean, my feet warming on a large rock in front of my mat. Sean lay on the mat beside me with his back turned up to the sun, but he finally glanced up at me and said, "Good for you, Bec." I was glad for the recognition.

We all have something we're willing to strip when we're far from home, and forgoing vanity and pride for a couple of hours on the beach was my way of being European.

That night, Sean was in bed early after dinner. At this hotel, he'd insisted on separate beds since we could afford them, and I was a little resentful at the passive-aggressive point he was making. We were in beautiful and romantic France, for chrissake. "Lighten up and live in the moment," I wanted to tell him. But besides all of that, I was still jet-lagged and couldn't sleep, so I walked to the beach. I was wearing a tank top and the same blue bandanna as my own badge of honor. The ocean was even more beautiful at night. The quiet of the city made the sound of the waves against the shore more clear and the rocks were still warm from the afternoon sun.

Then I heard a male voice, French accent. "You want a beer? You are American, no?" How can they always tell? I thought. I looked to my right to see a short, muscular Frenchman sitting a few rocks away smoking a cigarette. "Pretty hair," he said, pointing to my bandanna, and he smiled.

I hesitated and looked around for other people just in case. Several groups were sitting and talking on the rocks around us, smoking various things. My sense of adventure, spurred on by the confidence I felt after the hairless afternoon on the beach, took over and I walked over and took the beer and the cigarette he offered. We introduced ourselves. His name was Paolo. Eventually, I took another beer and another cigarette and told him about my vacation so far and how I dreaded going home. In broken English, Paolo told me he was originally from outside of Paris and wanted to be a writer.

He told me he was one of the vendors selling beer and bread on the beach in the afternoons, especially during tourist season. August was the month most French folks went on "holiday" and much of the country shut down. That's why we came here in August, I said. August is his best month, he said.

He said he saw me that afternoon sitting baldly in front of the world and watched me every time he passed. His drunken European eyes looked me up and down and the hair on the back of my neck stood straight up. I laughed and told him he was a big fat liar because I knew he really was looking at the bare breasts and forbidden nipples dotting the beach.

I'll never forget what he said next. "All the time I see breasts, lots of breasts, breasts, breasts. Not different," he said. "Today I saw different." Then he kissed me. It was one of those nice first kisses with no motives or expectations behind it. My neck and my ears tingled. He pulled away and I saw his eyes—a beautiful light brown, a little dilated from whatever he was smoking earlier. I didn't move away for a few seconds longer than was probably comfortable for him, wanting to commit this simple, intimate moment to memory. It may be the last one for a while and I wanted to be able to recall it the next time I missed kissing.

During the kiss my tank top strap had moved so he spotted the catheter. "You are okay?" I nodded. He didn't ask what the bump was. All I wanted to do was kiss him again, so I leaned in and he obliged.

We drank beer, smoked cigarettes and talked in broken English for hours that night, kissing periodically. Honestly, I enjoyed the kissing more than the talking. A couple of times he tried to touch places he hadn't gotten to see that afternoon on the beach, and I pushed his hand away, but it was nice of him to try.

I don't know if it was a line Paolo had spewed dozens of times in a tourist city strewn with women, or if we were just drunk, or if I remember it as much more romantic than it was. Maybe he was trying to get lucky, and he was disappointed when I didn't invite him to my room at the end of the night. You know, I really don't care.

He made me feel like it wasn't strange that I was sitting on the beach in Nice, France, kissing a Frenchman who thought my bald head was beautiful. I needed that more than I needed the vacation.

chapter ten

we are how we live

You are what you eat. We are only as good as the people and things with which we surround ourselves. It's always the last thing you eat, isn't it?

These are scary words for anyone who is sick. It makes you look at all you've done to and with your body—eaten too much grease, drunk too much vodka, bummed too many cigarettes at bars during college, breathed in too deeply when stuck in traffic jams or spraying hairspray, eaten way too many frozen or ready-made noodle dinners at the end of the month.

Once, my mother went to that scary place of speculation, doubt, and regret. She asked my doctor what he thought might have caused my cancer. He didn't have a confident answer, which is understandable, considering the world's top scientists can't provide a definitive answer to that question. It was surely an excruciatingly frustrating question for him and one he probably heard from every patient. I felt badly that my mother had asked, but was curious about what he'd come up with.

He tried, citing recent journal articles he'd read about studies investigating possible causes: inherently weak immune systems, exposure to large amounts of radiation including from microwaves and sun lamps, pesticides and preservatives, exposure to chemical fumes, a lack of quality air, even hair dye.

Suddenly, I wished she'd never asked, as I began to flash back to about a thousand apparently horrible things I'd done to myself. As a kid, I liked the smell of gasoline, kind of like enjoying new car smell, and never heeded the warnings that the fumes could be harmful. Impatiently waiting for popcorn or melting cheese, I used to stick my face in the window of the microwave and watch everything expanding and melting and heating when my parents weren't looking. In college I worked as a receptionist in a beauty salon and took full advantage of free cuts and colors, happily becoming the willing hair-color guinea pig for the newer stylists, as well as taking advantage of my free access to the tanning booths. I was told auburn hair looked fabulous with a tan. It was true.

I also thought about how my parents raised my sister and me on organic vegetables grown in our own garden, fed by bone meal and untainted by growth chemicals or pesticides. They made us eat all-natural, no-taste peanut butter and no-preservative Swiss and sharp cheddar cheese (my favorite) from a food co-op, and the only white bread, bologna, or American cheese we ever ate was at our grandparents' houses. We ate mostly free-range chickens and beef and pork, picked out by my father at family-owned farms. We tried dandelion leaf salad a couple of times after we weeded them out of our lawn rather than spread anything my sister, the dog, or I could ingest. We even twisted and boiled our own soft pretzels to avoid buying preservative-coated snack food at the grocery store.

My parents didn't consciously try to ward off something as perilous as cancer, but they were certainly trying to protect us from something.

When I was older and had some freedom to make some of my own food choices, I nearly always rebelled, of course, against these restrictions that none of my friends experienced—buying cheese puffs and cupcakes after school, splurging on burgers and pizza after the game. These were not forbidden activities in my household, but certainly not encouraged. I craved peanut butter that tasted good and bread with fewer than seven grains.

As an adult, that year or two before I got cancer, my older parents' garden was filled with trees and weeds, and their freezer filled with store-bought chickens. My father, the main gardener, eventually had his share of health problems that prohibited him from keeping up the organic traditions. But I began to return to the sensibility and eco-consciousness of my parents' influence. I still bought chunky peanut butter from the grocery, but I spent a few extra dollars on organic produce and Amish chickens and frequently ventured to the farmers' market on weekends. Maybe I was trying to make up for years of post-parental-control ingestion of pesticide-laden vegetables and powdered cheese products.

Several months later, I was sitting with my mother and my doctor, speculating for too long what could have caused the cancer, but eventually allowing a few moments of unabashed optimism to erase these scary thoughts. The chemotherapy was working. My oncologist happily reported that the grapefruit-sized tumor around the lymph nodes next to my right lung had shrunk to a golf ball—probably a benign one filled with rubbery fatty tissues. My blood counts, including the level that provided a clue about how much lymphoma remained in my system, were within normal ranges. My body was beginning to heal itself.

It's as if none of us could believe it, because when my doctor suggested a month of radiation to my chest, I was only a little resistant. I think we all just wanted to be sure it was gone. I chuckled at the irony that the chemotherapy had killed all my hair and most of the natural color from my skin, but I was about to voluntarily lie down in front of radiation beams to kill off whatever the tanning beds might have helped to multiply. Maybe I'd get some color while I was at it.

As we left, my mother brought up the list of possible causes my doctor offered a few months before—assuredly a precursor to a discussion about what I should do differently now that it was starting to go away. I awaited words like, "Now that we know what your body is capable of, what you're at risk for, let's not let this happen again." Nothing like your mother to bring up the elephant in the room to spoil the mood. She wasn't blaming me and, frankly, was saying what anyone would have at least wondered. She just wanted to be sure none of us would ever have to do this again—meaning she wanted to go on the record as unsupportive of tanning beds and processed cheese.

Honestly, I hadn't thought of the dreaded and still incomplete list of causes since the doctor's good news. To think of that would have meant days of dealing with doubt, regret, and guilt about moments in the past that closely resembled those cluttering the memories of the vast majority of people in the world not fighting cancer—all of which would have tainted the present moment filled with hope for the future. Besides, I think I knew I would have to start worrying about the future risks all too soon. For now, I just wanted to enjoy knowing I was going to be okay.

Two weeks later, curiosity, or the need for

self-punishment, got the better of me. I suddenly became aware of the survivors from my support group and the eccentricities they shared. More than once I saw one sitting, white knuckled, as I sipped a soda straight from the can. During the colder months, they began covering their faces with their coats when a car passed them in the parking lot. I knew they knew something I didn't, and I was almost sure I wanted to know what it was. So, I attended a seminar on nutrition and "life habit changes" held at the support center and recommended to cancer patients, but especially to survivors.

People who choose to be informed, even most of those who don't, are told repeatedly we're at higher risk for dozens of things—viruses, related cancers, after-effects from treatments, recurrences, side effects from medications—and one man was offering us things we could do about it. It was so exciting and depressing that, of course, most of my support group attended. We all sat up straight like good students, holding our coffees and notebooks attentively, waiting anxiously to be told how we should live, hoping for the burden of choices to be taken away from us somehow. As cancer patients and survivors, we obviously weren't able to handle them ourselves.

A very pale gentleman set up a table in front of the room while we watched. On the table he methodically placed a can of green beans, a frozen chicken, a glass of water, and a potted fern. Then the wizard spoke to us.

"Hello. Tonight I will address each of the elements you see on the table and tell you how they can either help keep you alive or help keep you unhealthy," the man said. He might have mentioned his name and degree or something, but I paid attention only to his dark and serious eyes. He obviously carried an anvil-like burden he felt he should share with as many people as possible. He did. During the next hour, the gentleman told us how each of the elements represented on the table could kill us if we didn't avoid it or take measures to clean it correctly.

Here's the full story according to him—an adapted version of his tirade, if you will:

He told us we should grow all our own organic fruits and vegetables. If we couldn't, we should scrub all produce with soap and water to remove pesticides, even if that produce was "organically grown" because we'd all be horrified if we knew how many companies lied on packaging. My parents' kindred spirit.

He spent a long time talking about store-bought poultry and pork that he told us was injected with hormones that promote the growth of cancer cells. Consequently, we were to cut all poultry and pork from our diet, unless decidedly free-range or Amish-raised, and why not beef while we're at it—no turkey at Thanksgiving, no ham at Easter, and absolutely no burgers on the Fourth of July. Besides, he said, it is almost impossible to find any beef that is pure enough, we'd be appalled at the horrific places these animals are killed in, and it's too hard on our kidneys and intestines and heart anyway.

I involuntarily grabbed at each of these organs as he named them and remembered one of my favorite episodes of *The Simpsons,* where a film narrated by Troy McClure was used to convince Lisa's class that dumb animals "would kill you and your family too if they had the chance" and all vegetarians were "Grade-A morons."

Our speaker took this assertion a step further and told us we should never buy frozen or ready-made foods or eat at restaurants because one has no control over the quality of the food, how they clean it, where they buy it, what they spray on it or inject into it. The same is true if you go to someone's home for dinner, I thought, remembering the cooking skills of some of my young friends who weren't exactly sure how long to cook the chicken, so we all just guessed that it was done when it turned brown.

Better to be safe than sorry, he said again, having said it several times already. I wanted to punch him because we all knew he meant to say that it was better to be safe than dead.

We were all a little too close to the concept to be able to laugh at his flippancy.

I watched the entire room became more and more pale as he continued. People began shifting in their seats and getting up for more coffee. No one left, but we all began unconsciously turning in our seats toward the door, hoping to be freed from this horrifying man soon. But none of this stopped him.

He continued by telling us that we should never drink tap water because it's filled with chemicals the government has added or not filtered out properly. If we have to drink tap water, let the faucet run at least twenty seconds first to be sure to clean all the cadmium off the pipes. I was afraid to ask what cadmium would do to me, but imagined inch-thick layers of scum and germs loosening and riding the water-park slide of pipes right into my glass and involuntarily gagged. However, he said, we should never trust bottled water because sometimes it's just filtered water from some plant in Colorado. What the hell are we supposed to drink, Mr. Horrible? I was about to ask. He obliged. We should drink only distilled water, he said. Yes, it's tasteless, but it won't kill you.

Finally, he got to the beautiful green fern at the end of the table. How could he possibly ruin the beauty of nature for us? I wondered. He didn't, exactly. Instead, he quoted studies that claimed the quality of air one breathes directly affects one's immune system, the lymph system—the organ in charge of fighting off the cancer and maintaining some sort of normalcy in the body while its poisoned, the same system that also had contained my cancer. Air is best if it's at least seventy percent oxygen, he said. If you live too close to a busy street, your air quality is unavoidably inferior. The only way to naturally clean it is to get some of these, he said, pointing dramatically to the fern. My apartment was right off of a busy street, and I'd never been able to keep a plant alive longer than three months. I was doomed, I thought, and I lowered my head in shame.

We all walked out of the seminar a bit dazed. We were given bodies that apparently couldn't withstand the environmental pressures of the modern age, but we had the awareness to take back control of our own environment and refuse to let the bad stuff in. It was almost too much responsibility. It was a lot of work and money. Not to mention, to make such sweeping changes was a constant, unrelenting reminder of one's physical weaknesses and risks. The grocery would be a worry-filled trip instead of a happy romp through the gustatory possibilities of life. Finding a reasonable and acceptable apartment in a busy street-filled city would become a stressful and lengthy endeavor. Watching others eat whatever they wanted would be an envious and bitter activity, causing a huge gap in my social life.

Some people walked purposefully, probably racing off to buy jugs of distilled water. But most of us walked slowly, gradually trying to decide if our current habits were a game of Russian roulette, or if this guy was full of it. The scary thing was, I'd heard similar assertions everywhere, from the woman behind me in line at the grocery store to *Nightline*. Even if I were completely healthy, these claims would be unrelenting.

On the way home I'd planned to go to the grocery store. I had nothing to eat for the week, so the trip was a necessity. But now I dreaded the usually fun trip.

As always, I went to the produce aisles first—my way of being sure I actually bought some instead of justifying at the end that I'd already spent too much. But this time, fear of waxed apples and pesticide-laden lettuce overtook me, and I walked straight through to the frozen section. Frozen french fries, my favorite, stared at me and I picked up the bag to read the ingredients. Polysyllabic words confronted me and I threw the bag down. The deli held some possibilities in that I saw some turkey with "no preservatives" and "all natural" stamped on the price signs, but then I questioned their truth in advertising and walked away.

Then I turned into the cheese aisle—a few years before the gourmet kiosks were found in the deli of every market in the country—with the blocks of cheddar and Swiss and provolone and crumbled bleu stretching along an entire aisle. It was my favorite aisle. I walked its length as I always did, perusing the labels and sales, smelling the newer brands to see if they were fragrant and turning the packages to check for any signs of mold. I was trying to decide between a nice mild cheddar to slice thinly for sandwiches and some soft water-packed mozzarella to cube into a salad. Maybe some olives too just 'cos, I thought, and my heart leapt with joy.

Suddenly, I remembered that I was supposed to be remembering something. Oh yeah, preservatives will kill me. I became embarrassed and looked around quickly to see if anyone had noticed me falter back to the dark side of casual food consumption, which had possibly caused all of this mess in the first place, and started to turn over the package of Colby I was holding to read the ingredients. Then, I stopped myself. What was I doing? Who's in control here, anyway?

I began to fume as I realized I'd given in to the fear instilled only minutes before by a perhaps-well-intentioned man with serious and sad eyes, not to mention a list of random and unproven causes sympathetically provided by my doctor, and completely chased away the contentment I felt just looking at foods I had enjoyed without wondering if they would give me warts or asthma or increase my risk of carpal tunnel syndrome or, in all seriousness, a recurrence of cancer. It made me ashamed and, frankly, pretty angry.

Right there I concluded that the gentleman leading the workshops was somewhere in between a knowledgeable person with important information to impart and a man with a bit of an

obsession. Undeniably, his words rang of some truth and reasonable caution, but they also rang of fear. Fear of chemicals and
impurities we all should be wary of? That was part of it. But I think
he was just afraid of truly living. I finally decided that the day I
become afraid to eat a good meal with friends at my favorite
restaurant, to enjoy an apple picked right from the tree, to refuse
water offered to me from a tap when I'm thirsty, is my saddest day.
Giving in to fear of the unknown for a few more fear-filled years
is not living, but merely a tedious, continuous exercise in avoiding
death.

So, like millions of people, I buy organic produce, if it's not
too overpriced, and try to eat less meat. I rarely drink water at
home that isn't, at least, out of a filtered pitcher. I adore walking
along the cheese aisle and always spend at least a little time
there, without reading the backs of the package, whether I buy
anything or not. And most of the time, I wash the store-bought pear
or nectarine I eat every day pretty well, but I sometimes forget and
just bite into it with wild abandon.

chapter eleven

first steps

"I'm sure this is hard for you to hear, especially because it looked so hopeful. I think we've got another fight on our hands."

There I was again. It was the same tone of voice—sympathy mixed with an air of distanced professionalism—only I was talking to my doctor by telephone, and the still-life was replaced with a nighttime skyline photograph hanging in the conference room next to my office.

I didn't say anything, but my doctor told me the chemotherapy and radiation therapy had not been as successful as we had originally thought.

Only three months before, I'd begun daily radiation treatments. A day or so after my twenty-third birthday the radiologists drew little right-angle marks with permanent marker onto my right middle chest and created my own personalized setting, a bright red square to match the marks, on a big machine in a small room. Five days a week I walked into that room, bared my chest, and lay down underneath the mammoth contraption where they covered my ovaries with a heavy black blanket, left the room, stood behind a protective wall, and told me to relax. I breathed in and closed my eyes, imagining lasers frying up the mass in my chest until it was nothing but a little sluglike piece of squirming rubber that only needed a shake of salt to shrink away to nothing. It didn't seem to work. When my radiation oncologist told me that the mass—a mass she admitted she wasn't positive was still cancerous—hadn't decreased in size any more and they wanted to do another month of daily treatments just to be sure, I insisted on a biopsy. Apparently, some of that patient empowerment I'd read about and talked about distantly at my support group was actually taking hold. Calmly yet defiantly, I told them I didn't want to put my body or my ovaries through more zapping if they weren't even sure the cancer was still in there. They argued, but when it came down to it, they couldn't force me under any contraption if I didn't want to be there.

My oncologists recommended a surgeon, and I went under a knife held by someone, perhaps the surgeon, who performed the biopsy happily, but also gave me an enormous amateurish scar, along the side of my right breast, that I still have. At the time it didn't matter because when they called me to tell me the piece of the tissue they'd taken was benign, I was elated. Sean was there with me when it happened. We celebrated calmly for about ten seconds. I wanted to hoot and holler, but Sean simply said, "I never expected anything less," and gave me a hug. I chose not to recognize the words and hooted inside. Within the hour we began discussing his planned move to New York City to pursue his acting career. He left in November.

My family and I enjoyed an unusually celebratory Thanksgiving. I'd started going to my support group only every few weeks. I was feeling less tired and actually began to recognize a normal life again.

Then in January I went in for a follow-up, largely unworried. After a very brief examination, my doctor didn't say anything and left the room. Then he tried to appear unconcerned as he told me he wanted me to stay for an ultrasound and a needle biopsy, "just to be thorough." Now he was telling me that while we were all fussing over my right chest, another tumor had engulfed the lymph nodes next to my right kidney. He wanted me to come in and discuss "next steps." I wanted to tell him that the most logical next step was to get the damn things out of me this time, but restrained my anger. Medicine is not an exact science, as much as we all want and need to believe that it is. I calmly agreed and hung up the phone. Then, my automatic pilot kicked in.

I took a few shallow gulps of air and in one sweep took care of everything I needed to take care of before actually dealing with this news. I walked out of the conference room and up to my boss to say I would be out of the office part of the next day for a doctor's appointment. "The cancer's back," was the only explanation I gave. His eyes went wide and he stopped his mouth from saying whatever was his initial reflex to say. Then he said to take the day and let him know what was going on when I could.

As I turned around, everyone who could see the boss's office had their eyes on me, wondering, but probably knowing, what was going on. They all continued to watch me as I picked up my bag and walked to the elevator.

Only my composure didn't quite make it to the parking lot. One floor down I stopped, not sure if I could find someplace to go if I started to lose it. And I just couldn't walk out into the revealing sun; it seemed too hopeful an image. I walked quickly down the hall and finally into the empty bathroom no one used. I couldn't catch my breath. I probably looked fairly calm, but my insides felt hysterical. I began pacing and looking for a place to sit that wasn't a toilet. I even walked to the middle of the room so if I fainted I wouldn't hit my head on anything porcelain, then laughed at myself nervously as I stood with both feet wide apart in the middle of the large cold room and my hands up ready to catch myself.

Despite the very determined stance, no clear words came to mind yet—nothing clear at all. All I could see were images of me in a hospital bed, me hooked up to dozens of clear tubes, me losing my hair again, me retching if I had to go back to the outpatient cancer unit and smell those chemicals one more time. I couldn't push these images out, and the effort made me more breathless. After what must have been only a few minutes, I realized I probably wouldn't be able to compose myself in a public bathroom and if anyone came in she would see me standing like a Ninja preparing for the dragon about to come out from one of the stalls. I needed to go home.

Outside it was crisp and cold. I'd forgotten my coat, so I swore loudly, raced to my car, and turned on the heater. I drove home, trying to figure out a way to tell my parents without letting them know how scared I was. Like me, they'd enjoyed a nice few months without the cancer—the uninvited member of the family. I hated this.

Once again, just like a year earlier at my first diagnosis, I sat alone in my apartment, breathless, refusing to call anyone yet, contemplating and questioning the next few months—what I would do, what I would have to do, what those around me would think and feel when I told them, even what I would look like if I was really, really sick. I imagined my mother's face becoming scrunched and red as she tried to hold back her tears. I saw my father put his hands into a pyramid and hold them to his mouth or chin while he tried to come up with an appropriate response that didn't reveal that his deepest and darkest fears were being realized. I imaged Sean sitting in line for an audition, not studying his lines, and feeling guilty for having dreams that took him so far away. I didn't like any of these images.

I was sad, angry, and scared—the expected emotions—but I think I sat so long because mostly I was confused. My emotions were used to being in turmoil recently, and my body was obviously well aware of what was going on. But this situation caused my brain to be hopelessly left behind.

A year and a half earlier, I had been a healthy twenty-two-year-old. I had graduated from journalism school with honors earned with sweat and tears and the sacrifice of the typical social life my fellow alumni remember fondly. Then, I happened upon my first professional position in a new city. I was enjoying my success, looking forward to all the careers and relationships before me. My biggest worry was that my car kept breaking down, and I'd have to save up for a new one eventually.

Now, I was a hairless, tired, pale twenty-three-year-old with a job I didn't like anymore, a hopelessly uncommitted romantic life with a man who'd just left for the East Coast, and a stack of endlessly complicated and constantly arriving medical bills. My smallest worry was saving for a new car because, even if I could get together enough for a down payment, I wasn't sure I'd be around long enough to pay the rest off. The last thing I wanted to do was stick my parents with car payments.

I know it doesn't sound like it, but I wasn't really feeling that sorry for myself—more just staring the sad reality in the face for the first time and hating it with every molecule in me. Is that the same as feeling sorry for myself?

There had to be a reason for it all, and I was determined to figure it out. Did I do something wrong the first time? Did I not fight hard enough? Change enough? Want recovery badly enough? Not have to sacrifice enough? Depend too greatly on people? Treat my support group with too much disdain for reminding me of what I could have been going through? What did the cancer want from me anyway?

My mind raced over the possible reasons and finally stopped abruptly. It seemed like such a non-health-related reason, but it made total sense. We all expected it. Just hear me out.

Since the sixth grade, I'd never accepted anything but the best from myself. I didn't just know that I could do better, I typically yelled at myself like a bad little league coach, calling myself names and asking what the hell I was thinking. This was not an external pressure. My parents expected my best from me, but they didn't push me to be more than I could be or scream at me from the sidelines to "quit being such a weenie." This was something I masochistically imposed upon myself.

In school, I always got the highest grade, and it used to bother me when I got an A-minus. It traveled with me to activities I claimed to enjoy as hobbies and extracurricular pursuits only. I avoided athletics besides running or whimsical games like dodge ball because I knew I couldn't win. I cried for two days after I lost the first-chair clarinet title to the boy I had a crush on in the eighth grade. He was shorter than me and practiced less than I did, but he was still better than me and I resented that. We also just happened to be "going with each other," and I was angry at myself because I had to fight the impulse to give in to him so he wouldn't stop liking me. It made me seethe that he won, even though every cell in me wanted to beat the pants off him. The anger and embarrassment of defeat made me burst into tears and run out of the room when the new placements were announced. Despite this, he remained my boyfriend, but I continued to be unhappy with second best.

In high school I was valedictorian of my graduating class, but only because the young man with the highest grade point average had been in the country only a year and was ineligible for the title. It ate me alive that I was secretly second, but I still beamed with pride as I walked down the stadium stairs and received my diploma first. No one had to know.

It didn't stop with measurable achievements like grades and band placements. Along with being friends and lovers and whatever else we decided we were or weren't that month, Sean and I were constantly, and often unconsciously, gripped in a battle for control of the relationship—a competition for who could tip the scales their way for even a moment. When I was still in my first round of chemotherapy and he was very busy with school and theatrical pursuits, I sometimes would suddenly become upset and need him greatly—forcing him to pay attention to me. This was the game I played with him on the days he wasn't playing similar games with me—pretending to be unattainable to make sure I kept coming back. In a relationship with love and passion, but without mutual understanding, this was the inevitable result.

For example, one evening I was feeling particularly restless about an upcoming treatment—sitting at home, shaking my foot like a junky desperate for a hit. Perhaps unconsciously, I decided to focus on one of the more swallowable issues in my life. So, I suddenly became remarkably upset, convinced I was being neglected and rejected by Sean. I became so agitated by this conjured crisis that I surprised him by showing up to his rehearsal—a good sixty miles from my home in Cincinnati.

When I found out he wasn't thrilled to see me and wasn't going to drop everything, including changing his plans to go to dinner with a group of friends that included a woman I knew he'd dated, I started crying and threatened to leave. Unintentionally, I think, my hat and scarf came off in the melodramatic mêlée in front of several of his friends, and I'm sure I looked pathetic and small, and probably pretty sick. Because it was a battle he wasn't willing to concede, he still didn't change his plans. He just walked me to my car, wiped away the tears, and watched me drive away.

But these were battle lines that were drawn vaguely. Had he won because he didn't let me manipulate him into spending time with me when I knew there were more important things for him to do? Had I truly won because he looked like an ass in front of his friends for not giving in? To the outsider, and to us a lot of the time, no one knew if anyone won our little melodramatic battles. No one understood why it was important. Truth is, we knew all along that only our relationships and our egos suffered. But both Sean and I were willful enough that we still really wanted to win. After each battle we went to extraordinary lengths to convince ourselves that we had.

Well, there was no gray area in this cancer thing. Everyone knew I hadn't won this battle. During my first fight, I carried these characteristically high expectations, as did everyone else. It didn't matter that I was fighting a life-threatening illness. I, along with everyone, didn't allow myself to view death as a possibility. My prognosis was good, after all, and I had a reputation of success to protect. Everyone else bought into the spiel and none of us accepted any excuses. For me, the possibility of dying was there, but I didn't admit it. And for everyone else, they didn't allow the possibility to cross their minds until the doctors told them it was time to.

It was finally time. It didn't escape any of us that if I failed to beat it this time around, I wouldn't get to pretend I did or say I did whether I did or not or whatever I needed to do to feel good about the outcome. If I lost, I died.

But maybe that was it. Maybe the previous harshness, the consistently and rigidly high expectations, were the problem. How many years of that kind of pressure can a body and mind endure, anyway? Maybe I was being given a gift—a glance into what it would have done to me long term if I didn't put a stop to it. Maybe I was being given a second chance—not only a chance to make myself healthy, but also to make peace with myself and with everyone's expectations, to make it okay to fail.

Furthermore, I'd have a chance to do it with everyone else behind me this time. Everyone would finally believe what I didn't have the guts to tell them the first time—this is serious business, no matter how positive the doctors are. They'd finally understand the emotional roller coaster I was riding and, maybe, want to get on it with me.

These realizations were the first steps. The first steps had to be followed by the next steps, but I didn't realize how uncomfortable they would feel.

chapter twelve

next steps

The day after my oncologist broke the news of my recurrence, my father came into town for my doctor's appointment. On the way to the hospital, we were quiet and seemingly calm. At that point we had no idea what we were in for and enjoyed not exactly knowing—feigning ignorance in need of imaginary bliss. This was our family's modus operandi. I'd seen it a hundred times, especially when it came to our own family traumas. When I was growing up, these usually arose when dealing with my sister—a woman whose adolescence and young adulthood was filled with depressions, suicide attempts, and hospitalizations. During each of these family traumas, it was the same m.o.—to different degrees, but always similar: When in crisis, pretend you're not, and you'll get through it fine. It worked like a charm, or so we convinced ourselves.

When I was twelve, I woke up one Friday morning at 4:30 to what I thought were the neighborhood cats yowling or fighting. I put my head under the pillow, but the noises persisted, and I sat up in the dark, forced more awake by one of the sounds that seemed distinctly human. Confused and half asleep, I looked out the window over our large yard and smelled the dampness of an early frost. I saw nothing, but I knew it was the source of some sort of animalistic moans that were now louder and more regular. I went into my sister's room, a little closer to the front of the yard where I thought they were coming from, to see if she'd heard them. I opened the door to a wide-open window and an empty bed. Rushing to the window, I instinctively knew what I would find, and my eyes didn't disappoint. I looked down and found my sister lying on the grass below, her legs twisted up beneath her, holding a long kitchen knife. Her voice was desperate and cracking from hours of tears and exhaustion in the cold air, but her words were clear. "Mom, Dad, please help me!" she was repeating, almost moaning from the pain. She was sixteen years old. I raced into my parents' room and flipped on the light, and my father woke up and heard the desperate cries. My mother went in to Lin's room, and I turned to see my father racing down the stairs.

I was shuffled to my room at that point, urged to get ready for school and get myself to the bus. My mother looked me in the eyes, and I could see hers were bloodshot, wide, and tearing from the fear and rage conjured by the mere thought of her child in pain—and the sadness that she hadn't heard her cries for help. I must have looked equally terrified, but all she could bring herself to say was that I needed to get dressed and get my books or I'd be late for school.

She didn't say, "Pretend everything is normal," but what else could all of that have meant? I sat in school all day not telling anyone what happened—half afraid that my friends would think I was a bad sister for just getting along with the day. Truth is, I didn't know how bad it was or even what had happened, what my parents did when they reached her, what Lin said when the ambulance arrived, or what reasons she gave for doing it. When I got home from school, my father took me to a friend's house where I stayed for dinner and into the evening. He drove me there in silence, and I knew not to ask any questions. When he picked me up, he finally talked, his voice shaking but his hands steady on the wheel, and told me that both of Lin's ankles were shattered and she'd said she wanted to kill herself by jumping out of the window and onto the knife she stole from the kitchen. "She's going to be in the hospital for a while." That's all he said for a long time, and then he asked me about school.

I never really learned the details, but I knew enough. I knew that my sister was deeply troubled and had chosen a very dramatic way to tell her family about it. I also knew that my parents rarely mentioned it, and I was urged to go to school and go to rehearsals like nothing was going on. I don't remember if I was allowed to visit her in the hospital, even though she was there for several weeks. I'm not sure I would have known what to say if I had. This was the first of several of my sister's episodes.

Ready or not, my family would face yet another life-threatening crisis. They would have to become more a part of the second fight against cancer than the first one—it was more serious this time. The modus operandi was still in place for car rides like these, when there was no good reason to think of the crisis at hand because there was nothing to be done that very moment. But somewhere between the years of my sister's crises and mine, the "rigid denial of reality in order to cope" had been worn down to a smoother "glossing over in order to get through what needs to be done." I think we were all a bit stronger and ready to cope with reality after the ordeals of Lin's youth and the expected family crises of the last ten years of our lives.

In Dr. Reinhardt's office Dad and I stared at CT scan films. "Glossing over" wasn't possible any more. The tumor in the lymph nodes in my chest was poisoned and baked by radiation down to a rubbery mass of benign tissue, but there was the other one, boldly smiling into the camera, hanging out around the lymph nodes next to my kidney. We found out the hard way that the cancer was stronger and much more cunning than anyone expected.

"Here are your choices for your next steps," the oncologist said. "We can do another round of a similar but more powerful chemotherapy like we tried before, perhaps some more radiation. It may kill it again, but then it may come back again."

Both Dad and I shook our heads involuntarily. It seemed weak and inconclusive.

"Option two, a bone marrow transplant, is a bit more risky, but I think your best option." My father and I were silent. It felt a little like an act of desperation, an attempt to grasp at the last few breaths of life. Physically, I felt better than I had in more than a year, and, at the time, bone marrow transplants weren't common treatments for people not approaching their final months.

Dr. Reinhardt told us of a doctor who ran a bone marrow transplant unit at a nearby hospital and gave us the rundown. I would be given two or three rounds of weeklong, continuous chemotherapy—just enough to get the cancer back into remission again. Then, if my bone marrow tested cancer-free as they expected, a surgeon would knock me out and extract a few bags full of the white milky stuff from my hips. Those bags would be treated with drugs and frozen.

Immediately, I'd go into the bone marrow transplant unit, a quarantined floor equipped with air filters and strict sanitation. My first week, they'd give me very high doses of chemotherapy, enough to clean me out better than a vat of Drano, killing any remaining cancer cells, any fast-growing cells, and my entire immune system. Then, they'd give me my juiced-up bone marrow back—an autologous bone marrow transplant—so that my immune system could rebuild from scratch. If that's the way it went, I'd be in the hospital at least a month, probably more. It would be a whole different story—including donor matching, months of waiting and hoping, and pills for the rest of my life—if I couldn't use my own bone marrow.

"There are risks, but chances are in your favor for coming out the other side not having to go through any of this ever again," he said. It sounded so dramatic: "coming out the other side." He understood something about what I was about to go through that I didn't.

The risk taker and impatient patient in me made up my mind somewhere around "killing any remaining cancer cells." But the thought of not going through any of this ever again had my mind reeling with hope. I didn't even have to look at Dad to know he agreed.

We left the office, relieved that a decision had been made. I was scheduled to go into the hospital four days later for my first week of chemotherapy. The most important "next steps" had been decided, we were moving forward, and I was ready to get back to business as usual. Modus operandi, learned at age twelve.

So, of course, the first thing I did was ask Dad to drive me to the grocery so I could pick up some dish soap and fruit. I was painfully low on toilet paper, too, I told him, and cat litter. Dad didn't say anything.

At lunch I rattled on about work and projects for the next week and next month. There was a nonprofit project I had my eye on, and a pro bono media project I'd been specially asked to head up on my own. I was sort of excited about something at work for the first time in a while, I told him. My father didn't say anything and just ate his turkey sandwich.

Maybe it was because he was curiously silent, but by the end of my one-sided conversation, I realized what I was doing. Pretend everything is the same, and it will be. Pretend there aren't enormous decisions and even more changes to be made, and they will go away—or at least you won't have to think about them. Staring into my father's now-older and wiser face, I knew he was waiting for me to realize it was not the time to bank on the family modus operandi.

The truth set in, but Dad didn't say it. He didn't say that there was no way I'd be able to see any work projects through, that there was no way I'd need more cat litter or soap or anything else for my apartment because I wouldn't be living there anymore. Dad didn't say lots of things, including that he knew that, because of the bone marrow transplant, I would be in the hospital for a week at a time, then possibly months. I would be weak and at high risk from communicable diseases and bacteria. My job, the apartment, the cat, the support group—it all had to go. Everything I'd worked toward in my brief independent life would all need to be put on hold, even more than the first time, so I could concentrate on getting well and be taken care of. He didn't say it, but I knew. My life would change and sacrifices would be made, but even though he didn't say it, I knew that my family would help take care of me.

It was the most difficult decision of my life, but it didn't take long to make. We knew that if the bone marrow transplant was going to happen, it had to happen before the cancer had time to get confident. Finally—and I wasn't happy about it—I asked my father if I could move home.

That afternoon I told my boss I'd be gone for a while and I hoped that my job would be waiting for me "on the other side." Then I started making plans to let my family face up to the challenge of taking care of me.

chapter thirteen

homesick

When I had my recurrence, I was twenty-three years old. I'd been living outside of my childhood home for most of the five years and, after college, living truly independently for almost two years. It doesn't sound like a long time, but those years were all I needed to know that I liked it.

Despite this, I swallowed hard as I put most of my things, including furniture, into storage. Immediately thereafter, I went into the hospital.

Once I got out, I went home with my parents, an hour's drive away from my former one-bedroom apartment. I expected to enter what had become of my old room—a foreign room filled with my parents' unused furniture, my parents' pictures on the walls, the same old nubby green carpet, and my parents' ironing board as the center piece.

When I walked in, I found an adult room of my own. The floor was brightly colored with my favorite Aztec rug, fresh from the cleaners. In front of the window was my own kitchen table holding several of my books. On the bed was the quilt I used as a bedspread, stains and all. My parents had taken things from my old apartment and made their laundry/storage/guest bedroom into my room again. It felt as much like home as possible, and I thanked them.

In front of the window they put my small farmhouse-style kitchen table with leaves and said it was for whatever I wanted to use it for. So, I went to the local art store and picked up some paints and pastels and some good paper. I bought a new journal and a set of pencils and pens. I set up the table, the lone piece of furniture I'd purchased myself, as an art station, a writing station, my own work area. I bought one small spiral notebook to keep next to my bed to write down my dreams for artistic inspiration. I knew I had to fill my time with something other than work and knew that puzzles or long walks to the park wouldn't be enough.

On the table I put a photo of the famous sculpture of the Greek goddess Nike of Samothrace, AKA the Winged Victory, that most likely pointed some Roman sea vessel to the horizon. I had taken the photo at the Louvre during my European adventure five months earlier. I stared at the headless, victorious, luminous creature, knowing there's no way that art in books has the same impact as the real thing. I wanted to embody the graceful strength of that sculpture—standing steadfastly as the wind blew her heavy wrap around her solid, womanly body. I thought maybe I could stand just as steadily if I tried hard enough and stared at my idol's picture long enough. So, the photo sat where I could see it from everywhere in the room.

Now, the room wasn't too hard, wasn't too soft. It was just right.

Overall, my parents and I treated each other as adults, at least in the beginning. I helped cook and clean up after dinner without being asked. I conformed to their household patterns . . . when I remembered. We chatted about politics and the local news at dinner. Dad let me have control of the television remote if I was there first. They let me sleep in whenever I needed to and respected my privacy, letting me sit uninterrupted in my room at my "work station" for hours, doodling, painting, or writing in my journal.

Out of consideration, I told them where I was going and when I expected to be home when I went out, but they never told me when I should be home. If I came in later than expected, my father was always still up watching TV or "reading because he couldn't sleep." He always asked how my evening went—a subtle pry into where I was—and I always told him because I knew he was concerned.

For me, the comforts of home were great: home-cooked meals every night, continuously clean bathrooms and floors (my retired father did the cleaning since it was risky for me to be around bacteria), days free to be introspective and build my strength for the transplant. I sat on the floor in front of my mother in the evenings while she gave me backrubs, and we watched television and gossiped about family friends, including controversial characters from their church congregation.

For them, they told me they were glad to have a job, some sort of role in their sick child's health and care. I also suspected they secretly enjoyed having some sort of control over how I lived and what I ate.

Still, no matter how hard we tried, after a while we all reverted at least a little to our parent-child patterns.

I rebelled a bit against their strict Monday and Thursday Laundry Day ritual, especially Thursdays as Sheet Changing Day (at home, I let the laundry build up for weeks, then spent all day at the Laundromat). After a few weeks, they began asking when I was going to be home, but only because I stopped offering the information so readily. When I saw the light on as I occasionally walked in the door at one o'clock in the morning, I breathed a sigh and rolled my eyes, dreading the forthcoming questions.

I started feeling guilty on Saturday mornings if I slept too long, but snapped when my mother woke me up to make sure I was okay. One morning I was awake, and she knocked softly to see if I was up yet. So, like any willful child, I pretended to be asleep. She sat on the edge of the bed for only about thirty seconds, and I could feel her eyes on my face, waiting for any signs of life. It's sad, but there was no way I was going to give her the satisfaction of knowing I was awake. I knew she'd want to talk, and, though I enjoyed our talks, I rebelled against the expectation. If I were in my own place, I thought, she wouldn't just walk in while I was sleeping. All I wanted was for her, my loving mother, to leave. Then, I avoided her for the next few hours because I felt guilty about it. I became convinced I'd traveled back in time to age fourteen.

The dinner conversations slowed, and we started watching the news while we ate. We all became a bit more agitated when we started bumping into each other, cleaning the dinner dishes. Then I started going out for dinner with my friend Fiona, an acquaintance from high school whom Joanie reintroduced me to not long before I moved back to Dayton. Joanie was touring a lot with the dance company she was stage managing, and Fiona, a surprisingly maternal woman at age twenty, gladly became the core of my social life and a refuge away from my parents' home. She got me out of the house and opened up her home as a comfortable, restful place to go even if I was tired. I started frequenting this refuge.

It got worse. After about three weeks the comforts-of-home honeymoon was over, and the house started to go sour. There are lots of reasons why:

As much as I may have craved quiet days filled with little but introspection and watercolors, too many of them became dull. That boredom, plus a marked lack of options, was bound to get to me.

As many of my own things as my parents could put into their laundry/storage/guest bedroom, it was still not my own place.

As quickly as we all agreed to take care of each other and be taken care of during this harrowing time, it was bound to feel like a burden to someone at some point. And, let's face it, too many days in a row in the same house post-childhood with your parents or with a willful adult offspring is going to start to drive you mad. Add a lot of unspoken fear and anxiety about whether one of us is going to live through the spring, and you've got yourself a cage filled with jungle cats pacing the boundaries.

We bickered. We snapped. We fought. We stormed out of rooms. Once we went an entire day without speaking. Another day the only time I spoke at all was to pick a fight with my Dad about the inferiority of their water filtration system. I think I called my own father—the man who received *Organic Gardening*, belonged to a food co-op for more than ten years, drank herbal remedies from the health food store, and gave a good percentage of his pension to environmental causes—"ignorant" because he didn't know that the cadmium found in household pipes increases the risk of cancer (it *may* increase the risk of *some* cancers). He just stared at me blankly when I told him that "without a proper filtering system we should always let the water run for at least twenty seconds before using it, *Dad.*"

Another day the bickering escalated to a few minutes where all three of us stood in the living room screaming at each other. I'd wandered into the living room and turned on the television to one of my favorite sitcoms. My father, usually the one in charge of the remote, wandered in soon after.

"That program isn't allowed in this house," he said matter-of-factly, sitting down hard after a long day. He was retired but his own health problems—fibromyalgia, chronic fatigue, high blood sugar to name a few—meant that maintaining his small antique business day to day was sometimes all he could handle. I smiled and looked toward him, expecting a smile in return after a moment of light-hearted sarcasm, but he was not smiling.

"Are you kidding?" I asked, still smiling and still not sure if he was suppressing one of his own.

"Absolutely not. That show is evil, and I refuse to watch it," he said.

"Evil? How is it evil? And why do you get to decide what we watch?" I said.

My father glared at me, and I almost caved simply because I hadn't seen that look since I broke curfew after Homecoming my senior year of high school. But I didn't. I probably should have because what happened next was like a flashback to that rebel-against-my-father time I thought I'd left behind.

"This is my house," he said in a louder voice. "That show is evil and is not allowed on this television. Furthermore, I'm embarrassed that you watch it." At that moment the show's main character entered the wide screen behind me and spouted a nugget of sarcasm at the oldest of her three children, a willful young lady known for a lot of yelling. "I would have hoped I'd instilled in you better judgment," he said, pulling out the disappointment card. I heard my mother get up from the kitchen table and walk into the room.

My jaw set, and my pride swelled, and I couldn't believe he was questioning my character because of my television viewing choices.

"The last time I checked, I was twenty-three years old and allowed to make my own adult decisions, including what I watch on television," I said with a similarly raised voice. "If you like, I'll go to the small television in the kitchen, but I refuse to accept that I am a bad person because I happen to like this show."

"I wouldn't want you to watch it out there either," he said. "It's an evil show. I don't want you to watch it."

"Evil?! What does that mean?! How can a sitcom be evil?!"

I said. I was yelling and standing up now, not believing what I was hearing.

Mom spoke up then. "You both are adults. There's got to be something more important to fight like this about," the voice of reason, even if voiced loudly, said. There was, but how could we fight about what was really bothering us? It was cancer. I might die. These were facts, and there was nothing to fight about. The state of the entertainment world and whether possession or ownership mandated control over the remote control were more palatable topics. The climax of the argument, occurring only a couple of minutes later, was when I stormed out in tears, and my father said that it was typical that I'd leave since I rarely stayed at home with them anymore. He sounded a little hurt as he said it.

I went out of town and back to the hospital in Cincinnati twice for weeklong rounds of chemotherapy in preparation for the bone marrow transplant. The second one rolled around, and a part of me welcomed the change, almost seeing it like a business trip—I wouldn't have chosen the destination, but it was a getaway from my parent's house, nonetheless. I spent my days leisurely reading the paper, watching evil sitcom reruns whenever I wanted, and visiting with in-town work friends. All I needed was a margarita with an umbrella in it instead of the poison cocktail the nurses were serving up, and I was on some sort of twisted sterilized vacation. Mom and Dad were scheduled to pick me up on Friday.

Then, something surprising happened. I was sitting awake at five o'clock on the third morning of my stay, staring at my drip bag. It was so quiet and stark white in the room and there were no chairs or cushions or fabric anywhere—just space. Nothing on the walls except a crucifix. No smells but the metallic chemo-drug aroma seeping through my pores. No light but the fluorescent glow from the bathroom. I started to get anxious without color or warmth, and my feet and hands became cold.

I closed my eyes, and the first people I saw were my parents. Not the disappointed face of my father or the watery-eyed concerned looks of my mother, but the parents who'd taken care of me and raised me with care and love. The parents who bought me the teddy bear I asked for over and over before my fifth Christmas and set it underneath the only lamp that was on in the living room so when we descended on Christmas morning it would be the first thing I saw. The parents who let me cuddle with them when I had nightmares and attended each and every one of my concerts and theatrical productions and recitals from the time I was seven years old. The parents who taught me that I could be anything I wanted to be and encouraged me to try different things until I found out what made me happy. The parents who instilled in me the individualism that caused my willfulness and much of their exasperation, but who never truly faulted me for it. The parents I kissed every night before bed, whether we were getting along that day or not.

Suddenly, I felt like I was desperately lonely at summer camp and wanted just to see my Mommy and Daddy, and everything would be better. Suddenly, I was so glad I had someone to pick me up and take me home and tuck me in. I was so glad I didn't have to go through any of this as alone as I felt in that cold, bare room. Who would've guessed that my family home could be at once a prisonlike reminder of my current restrictions and a womb-like protected place—the only place that could make me feel better. I fought it, but I started to cry anyway. I did manage to resist the urge to shove my thumb in my mouth and buzz the nurse to bring me my blankie.

My parents picked me up on Friday, and I was glad to see that we all had the same relieved look on our faces when they walked in. I tried to remember how I looked when I became consciously and fiercely independent, but it didn't work. I must've looked like a scared little child because my Mom uncharacteristically hugged me tight for a long time, and my hands and feet started to warm up.

When we returned from the hospital, my parents and I kept our toes dipped in those comfortable roles—the rebellious fourteen-year-old and the overprotective parents—but we tried not to. And we started not to hate each other for what our cohabitation was putting us through. That's all we could ask.

chapter fourteen

is dad always right?

My first box of sanitary napkins was enormous. Because I was a beginner, my mother decided I should have the blocklike Modess napkins you wear secured to a white elastic g-string that constantly reminded me I was HAVING MY PERIOD. Only nine years out from wearing diapers, I couldn't help wondering why this contraption was supposed to make me feel like a woman. There was no miracle or blossoming. It just seemed inconvenient and messy.

Mom tried to convince me how amazing it all was—that a whole new part of my life, a whole new set of opportunities, was starting, including the possibility of children someday. I was eleven years old. I was far from what I thought a woman was, and even farther from any drive toward motherhood. It was something I'd understand some day, my mother said. But for now, just watch for it every month and be sure to tell her or my teacher if it ever looked too red.

At first, there were few enough periods to count—"I'm on my seventh. How many have you had?" But we all lose count eventually, and the novelty wears off, especially the second or third time it arrives unexpectedly in the middle of a spelling test or under your choir robe at church. It finally started making me angry. Why should I have to deal with something I don't even use? Sitting on the toilet one morning wiping red streaks off of my leg, I wished it away.

Eight years later, at age nineteen, I sat on the toilet wishing for those red streaks. A pregnancy test sat next to me unopened, and I thought if I took back the preadolescent wish earnestly enough, my body would just agree and it could remain that way. By then I'd discovered the early miracles of womanhood—with the exception of fertility, the most-feared "side effect" of young sexually active women. The year before, I waited in the lobby while a good friend of mine lay directly above me with a vacuum twisted into her cervix. Then I sat in her living room while she cried and got up every half an hour to change the Modess pads they gave her at the clinic. I didn't watch the abortion, but just seeing the crudely cloaked sadness in her eyes when she stuck out her chin and said a little too zealously she knew she'd done the right thing and that she didn't have any regrets . . . The experience changed my mind about whether I could go through the same thing. I believe in a woman's right to choose the direction of her own life and her fetus's, but what a painful choice to have to make.

I knew if the red streaks didn't come, I'd be a mother and forever tied to a lovely but uncommitted half-boyfriend who thought more about his future plans to move away and pursue his career on the stage than about spending any of that future with me. I shook with fear and regret and tears. The streaks eventually came, and a month later I started taking the birth control pill.

I didn't go off the pill until my oncologist

said they would make the veins in my legs clot when I was in the hospital for my bone marrow transplant. Before that, he'd wanted me to keep using birth control because it was very important I didn't get pregnant during chemotherapy—"for both the mother and the child." Basically, they didn't want the mess of having to perform an abortion while I was still in chemo, and they knew I couldn't continue with chemo if a child was involved. My cancer not only determined most of my life (and death) decisions those days, but it also made my family planning decisions for me. My period still came like clockwork, but I was never sure if it was because of the pills or because nothing had been damaged from the nine months of treatments preceding this.

At the final doctor's appointment before my bone marrow transplant, the nurse practitioner, my father, and I went through a huge three-ring binder of information detailing everything that would happen, day by day, what drugs would be in my system for how long and what they would do for me, every possible side effect or health problem I'd ever be exposed to as a result. It told me that I shouldn't swim in a "body of water that is not chemically treated" for at least a year. "Damn, I'll have to cancel my skinny dipping trip," I said. Nurse Julie smiled, but my father only cleared his throat and rolled his eyes at me.

I got goose bumps after the two pages describing shingles, the very painful adult version of chicken pox, to which I would be susceptible, or the section describing the possibility that a casing around my heart could begin to materialize, or the part about my lungs hardening. It almost didn't register when a tiny little paragraph on page 15, subtitled Reproductive Changes, described how the highly dangerous doses of chemotherapy I'd be taking would most likely "disable" my ovaries, causing them to shut down. I paused for a moment, looked at the nurse practitioner. "Menopause? This is going to happen?" I asked.

"It's most likely," she said. "You should prepare yourself for it." I had only a week before the transplant and more things that I needed to "prepare for" than I could count. I heard my father swallow hard.

"Can anything be done?" I said.

"Well, you can harvest some of your eggs, but we don't know if they're still good after your earlier treatments . . ."

"How long would that take?"

"A while," she said with a big sigh. "We'd have to cancel your appointments and reschedule the transplant. You'd lose your spot in the study." That last sentence was said with that skeptical tone that swung upward to make sure I knew she didn't advise waiting. She knew I was the last of my oncologist's transplant patients allowed in a study of an immuno-stimulant that supposedly increased effectiveness of the transplant and sped recovery time. It was one of the reasons I chose this doctor.

I couldn't believe this was the first time I was hearing this. I couldn't believe I didn't realize or investigate this on my own. Menopause hadn't even crossed my mind, let alone preparing to have children. I was single, financially unstable, living with my parents, only a year into a career I didn't voluntarily choose, not nearly mature yet to handle the responsibility of a child, and, oh yeah, I'd had a recurrence of cancer and chosen the strongest treatment available so I'd be sure to see my twenty-fourth birthday. The pitter-patter of tiny feet wasn't really at the forefront of my mind just then.

My mother started menopause at least five years before she was officially barren. My grandmother's had taken nearly ten years. It had taken me t-minus fourteen months to officially become useless in the procreation of my species, not to mention my family or my future husband's family. My only hope for salvaging my dignity at this point was if I'd broken some sort of record. And I'm sure Guinness didn't have those times.

After this stream-of-consciousness-motivated pause, I abandoned my original line of questioning with some comment to my father about how I knew this was a possibility since my first day of chemo, and turned the page. There was no way I was going to delay my treatments and let the cancer take even more hold. That, at that moment, was my first priority.

In the car, I tried to sound as nonchalant as possible when I said, "Well, I guess no biological grandchildren for you, Dad."

He said, "No, but you'll be alive."

When I moved to New York a few years

later, then to Philadelphia, then to Los Angeles, I had to find a new general practitioner and a new gynecologist every time. Each time, the same questions about my health history and my reproductive practices were asked. Each time I gripped the side of the examining table when they said, surprised and dismayed, "Didn't you harvest any of your eggs?" Or, they assume ignorance and go into a very patronizing explanation: "You know, we can harvest eggs these days and freeze them until you're ready. . . ." I guess they think doctors in Ohio don't know nuthin' 'bout such thangs. Or that it's helpful to mention these possibilities several years after they were actual possibilities—just to show they know more than me.

It doesn't matter why they say it. Each and every time I feel about as smart as a litchi nut and say, "No, but . . ." and tell my side of the story. But by then it doesn't matter. They look at me with a look of disappointment and disdain, but eventually decide I'm worthy of their healthcare services after all.

As I'm putting on my clothes, the anger and humiliation starts to well up like a backed-up garbage disposal. By the time I'm backing out of my parking space, the disposal in my imagination is grinding down their perfectly manicured, well-trained hands and hoping their fertility drugs make them spawn sextuplets. It's so easy to be angry at my doctors—they at once saved my life and damaged it in ways that were out of anyone's control. It's the anger at myself that visits me more often.

For several years following the transplant,
my father's "But you'll be alive" comment rang more true to me than anyone else's justifications. Just like him, it was very logical, based on a farmer's rules of survival tactics, and it made so much sense. But I also knew that several heavy invisible knives were poised and shaking over me, waiting for the signs of maternal instincts to strike: 1) the first time I looked at a pregnant woman and didn't immediately become preoccupied with how she could sit comfortably; 2) the first time I realized I was spending more time in the children's books than the new fiction section; 3) the second or third time I held a baby for dear life when his mother wanted him back; or 4) my worst fear, fell in love with a man who wanted children more than a career.

With each of these, I knew one of the Infertility Knives would stab. Of course, they couldn't all dribble down steadily every couple of years or so; they hit me all at once, starting with the last.

Isn't that usually how it happens? Many women say they don't want children, or they don't want them right now, but that can all change in a matter of weeks should the right "father" come around. The same can happen to men. If you can't imagine yourself with a family, it's usually because the other leader of the pack, the perfect coconspirator, the right partner, hasn't appeared yet to make the scenario realistic. Once he or she appears, the primal instincts kick in and the cravings for a family become overwhelming. It's just the way it is, no matter what your thirty-six-year-old heterosexual single friend says.

So, at twenty-eight, my worst fear came true. I found myself lying next to a man I adored, a man whose character and desires cried out for fatherhood. Children gravitated to his strong frame, warm smile, and silly faces. His smooth, low, radio-ready voice commanded attention and respect without making them afraid. He talked to kids like people, not toys or entertainment, and he taught them to be good sports or good neighbors simply in how he asked them questions. He even joked around with the waiters at pizza parlors to make them laugh. He had that rare healthy quality I feared the most—no trepidation whatsoever about commitment or settling down. To him, family and relationships are the most important things in life. Before we were dating, a mutual friend described him to me as "a family man without a family," one of the most insightful descriptions I'd ever heard.

This man I loved was built from the ground up for fatherhood. He just had the misfortune of falling in love with a woman unable to give him biological children. But the worst part is he's never considered it a misfortune. "We'll adopt, my love," he's said over and over. "A family is a family, and I can love a child that doesn't have your eyes and my wavy hair just as much as one who does." I've never stopped feeling guilty about that.

Despite this man being so good for me, I married him. Even so, night after night I lay next to him, unable to provide him the one thing that would make him most happy in this world. Night after night I vow to myself to make him so happy that biological children won't ever matter. I've been lucky so far in that making him happy hasn't been that difficult.

Sometimes I hate my doctors and sometimes I hate myself. Sometimes I'm unspeakably grateful just to have someone who adores me without any reservations. Sometimes I resent him for being better than most people—for not wanting a career or freedom more than a family, for wanting me to be happier than he is, for giving me a reason to want to be pregnant.

Often I can't see straight for the irony and regret that my first priority when I was twenty-three, glancing between my nurse practitioner and the words "premature menopause" in a booklet, was living long enough to meet someone who would want to have children with me. I didn't stop to consider what it would feel like if I couldn't.

But at least I'm alive. Right, Dad?

chapter fifteen

house of prayer

When my parents and I arrived at the hospital the morning of my bone marrow transplant, my parents' current church pastor was in the waiting room. I had gone to church once since returning to Dayton and only because my parents urged me to say hello to all of the people who asked about me every Sunday. I enjoyed singing the hymns and talking with old friends, but other than that it was a fairly meaningless experience for me. Nevertheless, the pastor had specifically asked if she could be there, and I agreed, knowing that it would be a comfort to everyone except me, but wouldn't really bother me either. I'd realized long before that day that the people around me needed just as much comforting as I did—sometimes more because they didn't have the day-to-day realness of the disease to break it down into small swallowable pieces.

I wasn't always able to comfort them, nor them able to accept it, but sometimes I could "facilitate." This was one of those times.

The pastor followed us until moments before I went into the operating room. She began to pray, and I bowed my head politely, but I really just wanted to be quiet and sit with my parents. I'd already done my own form of praying—"peace-finding" is what I'd come to call it—the day before. She asked if I had anything to say or ask, or if I wanted to pray on my own behalf. I thanked her for coming out and for praying well enough for us all. My parents smiled, and I knew I'd done the right thing.

I was scheduled for two procedures that day. The surgeon would flip me over, drill into my hips, and extract two large bags full of bone marrow; and then he would flip me back over and insert a triple lumen catheter—a tube with three arms that made it possible to hook me up to three fluids simultaneously, and that led directly to an artery near my heart. Their goal was not to have to access the port-o-catheter already under the skin of my left chest so I wouldn't have a constant needle stick. I was happy for the stickless option, but not happy that I wasn't able to keep to my nothing-swinging-outside-the-body resolution. There now were four ways to access my bodily fluids, and I hoped there'd be no cause for more than that.

I was grateful that at least I was not awake this time for the bone marrow harvest. The previous three times they'd extracted the stuff—yes, three—I heard and felt every second. Sure, they gave me a local anesthetic, but no matter how numb you are, you're going to cringe when a doctor inserts a wide syringe into the bone he just ground into, huffing and puffing with a mammoth needle they didn't bring out until after you'd flipped over and the happy gas has started to take effect, and then suddenly pulls the enormous plunger to suck out most of your hip. After each test I walked around for days with an ice pack strapped to my jeans. There are no profound growth experiences that come from that procedure, and it was by far the most unpleasant physical experience of the illness. Yes, even more than losing my hair.

A few days after these preliminary events, I would receive several days' worth of extremely high-dose chemotherapy and then would receive my bone marrow back to help my immune system recover. That's the short version of the story.

I woke up in a large triangle-shaped room

with tall, wide windows to my left and a curtained-off area to my right. I already had all three lines of the catheter going with various fluids, and I desperately had to go to the bathroom—a sensation that rarely left me during the entire stay. After straining to find a door besides the two consecutive ones leading to the hallway, I realized the curtained-off area was my bathroom. When I tried to laugh about it with a nurse later, saying at least I didn't have to wheel myself down the hallway to the "community loo," she explained with a serious whisper that it was a bathroom I could use "even if I was very weak." It also would be the most privacy I'd get for a long time, so I swallowed what little modesty I had left, decided not to joke around with the same nurse again, and accepted it.

My room was at the end of the hall, the corner room, so I was able to see down most of the ward. I could see into one woman's room where she had a full stereo setup and what looked like work files spread across her bed. She looked up and caught me staring. The room next to mine was completely dark except for one steady light on the headboard, and when a nurse came in to check the patient's IV I raised my head to try to see the mystery patient with the additional light from the hallway. No luck.

Right outside my room was a small nurse's station and the entryway where guests scrubbed their hands. The entire ward needed to be protected from all bacteria so we didn't get sick. Until my immune system was able to handle the outside world, I'd be confined to the special unit with filtered air only, no flowers or plants, no raw produce, no dairy products, no unsterilized hands, and absolutely no ventures outside the unit.

My parents came into view so I sat up, and I could feel the anesthesia was starting to wear off. I'd almost forgotten they'd just drilled into my hips like oil reservoirs, and it felt like they'd used equipment just as large.

My parents had bags full of things I'd packed specially for the stay. We pulled out a tall stack of comfortable cotton clothing, including my favorite pair of oversized pajamas. In the other bag was a photo album containing pictures of friends, pets, family, cities, and artwork—reasons to get the hell out of the hospital. I pulled out a separate framed photo of Sean and me, and one of Joanie and Fiona. The bag also contained my sketchbook, a journal, art supplies, enough books for a year, a stuffed rabbit from Sean, and a silk bag containing healing crystals from my sister and her family. I set them all up next to my bed so they were the first things I'd see every morning.

Most importantly, in the bag was a photo I'd taken and had developed the day before. I'd walked around a local nature preserve all day with a camera, snapping tree buds and glistening water, watching the gray and brown starting to let the green show through a little. It was still the final days of winter, but everything was bulging and ready for spring. For the first time since autumn the sun was a little warmer than the crispness in the air was cold.

The creek that runs down the middle of the preserve was overgrown with bushes and dangling tree limbs most of the year. Without the fullness of leaves, they leaned in like arches over the creak, leaving a thin strip of blue sky above it to remind you you're outside.

It was in this spot that I did the "peace-finding" I recalled while the pastor prayed before I went in to surgery. I looked down the tree branch archway that narrowed and curved away from me like my own small cathedral and spoke with myself, preparing.

I'd spent a lot of time with myself in the roller coaster that was the past year—short spurts of mania, a couple of hours at a time spewing thoughts onto pieces of paper in words or shapes, rare long pieces of silent time intentionally filled with nothing but a strange peace. Not once did I ever really talk with myself directly. I listened and tried to interpret what was going on in there. I even tried to reason with it passive-aggressively—claim control over the uncontrollable entities swirling around. But I never took my words inward and had a conversation with myself. Honestly, I thought it wouldn't do any good. And I was afraid someone would start to talk back.

When I was a senior in high school, my father once said, in a conversation with a long-time friend over kielbasa and the best sauerkraut I'd ever tasted, that it was okay if my sister and I didn't believe in the same things my parents did—God, the divinity of Christ, everlasting forgiveness—as long as we believed in something. He simply wanted what most parents want: a basic moral code and, layered on top of that, a healthy spiritual life. Because of that, I knew I always would. So, I took his words to heart and decided then and there I would believe in the things I always knew to the core of my soul—the basic goodness of humanity, the sanctity of nature, the everlasting cycle of life, and the strength of the human spirit, including mine.

The rest of all the God and Christ stuff was TBD—to be determined. But because I was raised in a church setting, I still felt the need to pray in some way. To accomplish this sense of simultaneous peace and connection I tried lots of different things—self-guided meditation, Tai Chi, extremely liberal congregations of various faiths. I even involuntarily began to pray once, but stopped when I realized I had no concept of who I was praying to.

It was during a lone hiking trip—not my first hike taken alone
and by far the preferred approach—that I realized its meaning. I
was standing on a rock in the Hocking Hills, a beautiful and mysterious state park filled with Jurassic-looking caves, caverns, and
long curvy waterfalls in southeast Ohio. It was a spring Sunday,
and I was avoiding writing a fiction piece for a senior-year journalism class at Ohio University I was too intimidated by to enjoy
or learn from. I knew to my core that it was an important class for
me, but I couldn't bring myself to write anything good. I was failing, and we all know how I deal with that. So, I had to get away
and clear my head.

I kept moving until I found myself completely alone. I was surrounded by water, with a waterfall cascading down at least three
stories in front of me, noisy, scratchy sandstone beneath my feet,
and green-covered cliffs to my right and left. There were no people, no plastic wrappers, no empty sheets of paper, and no noises except the water and a few seconds of wind-rustled trees at the
top of the cliffs. They were talking to me. The blowing trees were
so green and tall above me that I knew they were more powerful
than I could ever be, and I believed that the breezy sounds were
meant only for me because I could understand them. I suddenly
felt very humbled and small, in the best sense of the word, but still
an essential part of this microcosm of the cycle of life I'd entered
temporarily. I thought this must be the bliss that a devoutly religious
person feels when she prays.

I was completely at peace for the first time in months. I can't
explain why, but I knew I needed to be sure all of me appreciated it. One by one, I woke up each of my major parts—the brain,
the body, and the soul—and made them look at such a beautiful
sight. My brain relaxed, my body tingled with every breeze, and
my soul smiled. I'm not sure, but I think they all thanked me.

After that day, I took hikes more frequently when it was warm
outside, and during the winter I craved them desperately. It sounds

crazy, but when I was unable to see trees out my window, I noticed. I became anxious and spacey. I craved the outdoors, especially trees.

So, the day before the most important and frightening day of my life so far, I understandably needed some nature. My mind raced and babbled loudly about every negative thing it could come up with—doubts about the unknown, doubts about my strength to get through it alone, regret about all I hadn't done with my life yet. When I found myself covering my head with a pillow and scrunching my face up to try to make it stop, I decided it was time for a nature-initiated conversation with the one person who could get me through all of this. The nature preserve, right at the mouth of the creek covered with a golden brown archway, became the place to do it.

I stood in the middle of the creek on a rock and tried to clear my head. I could hear a breeze starting in the trees behind me like a ghost sneaking up on me, and then I saw the trees before me shake easily one by one all the way up the aisle. I focused on the barely perceptible noise it made and the ancient smell of wet earth on the banks beside me. My face began to unscrunch, and I could feel my legs relax a little. A few minutes later, the weight of my head and my feet lightened, and I involuntarily rose up on to my toes.

Then I spoke. I told my body I'd do everything I could to help it recover if it just worked a little harder than normal the next few months. Stealing from every boss I'd ever had, I asked it to "take one for the team" and promised it would reap the rewards someday.

I told my mind to be quiet for a change, not babble so much, stay focused and let my intuition rule the turf for a while. It grunted a little bit, but reluctantly agreed to try. Those were the easy ones.

My soul and I were silent for a long time, and we both looked up at the quivering trees above. Finally, it spoke first. It said it was going to lie low for a while, be silent and strong. Like a good friend, it said it would be around when I needed it. I knew it was the only part of me guaranteed to survive the coming months, and I was okay with that. It was a good soul, I thought.

The short talk over with, all I could think to do was walk. I couldn't sit or go home or go anywhere near the hospital yet, so I started down the watery aisle in the middle of my outdoor cathedral, trudging up and over rocks through the chilly late winter runoff. The aisle continued up past the creek bend and to the edge of the preserve, and I kept walking. I passed a couple of deserted cabins with no signs of life. Was I looking for something? I don't think so. I think I was just compelled to move forward.

As I continued, I frightened a few bright green water frogs that disappeared under rocks, and then I happened upon several curious hikers. I even hunched over and walked through a tunnel under a country road, but never did find the end of the cathedral aisle. Only a fence stopped me, but the creek continued on past my line of sight. I wasn't exactly sure what that meant, but I took it as a good sign.

As I stood there, trying to peer up the rest of the creek bed, I finally found my peace again. I decided that no matter what happened, I was going to be okay. I decided to have faith in the cycle of life I'd added to my personal doctrine years before and try to relax. Sounds easy, right? It wasn't.

Besides all of this, I discovered some other things I knew as I walked down that watery aisle, things I should have added to my own personal doctrine but didn't have the guts. I never thought I'd have to think about what it would be like to be dead at twenty-three.

Therefore, there are some additions, perhaps a little more practical, to add to my previously very nebulous personal doctrine:

1. I would be disappointed not to see the nature preserve again when it was dripping with green.

2. Even in the after-life I would crave big hearty meals with wine.

3. I would be sad not to have the chance to get married or have children or go to Greece or a thousand other things.

4. I would miss my family and friends, including the ones I hadn't met yet, more than I could express.

5. I didn't want to die, and I believed that this trip to the hospital wasn't my time to die. But if I did, it was okay. I wasn't afraid of what might or might not be waiting for me, and I was pretty sure my soul was ready for the adventure.

In my hospital room, the photo of my tree-arched cathedral sat on a little cart inside the only place that seemed appropriate, the only private place in the whole room. Occasionally, a guest who couldn't make it to the public bathroom outside the ward commented on it. I just told them they were relieving themselves in my "house of prayer." Not one person realized it wasn't a joke.

chapter sixteen

what was the question?

Life was simple in the bone marrow trans-
plant ward. Because we were isolated from so many of the joys
and hazards of the outside world for what we hoped was a temporary amount of time, we all, even the nurses, were stripped of
the rules and expectations of that world. We weren't exactly
marooned, and we weren't even close to putting paint on our
faces and wielding conch shells, but we did appear to welcome
the ward's separationist approach. It was the way of the land in
which we lived, and we conformed.

In many ways, I was grateful for the resulting lack of required
decision making. When I was in control of my life, I spent every
day making hundreds of choices—what to say, what to wear,
what to eat, what to do, how to pay for it. In fact, with the introduction of the cancer in my life, I found myself in a constant and
tiring battle to maintain these choices and civil liberties. In the hospital, you have no choice but to let that battle be fought almost outside of yourself. You have little choice about anything.

You can always rebel against the doctors and nurses and choose not to "get some rest" or "take it one day at a time," but why would you? It's so much easier for everyone if you play along. So, I surprisingly and willingly—and temporarily—gave over the control I previously obsessed about. Once I did, I discovered that being freed of that self-imposed burden, allowing someone or something else to take over, worked.

But it wasn't always easy to keep up this freedom from the burdens of expectations and words. I had to do a little acting, and after a while, it became reflex. Whenever a question was posed, I played dumb or distracted: "What? Are you asking me? What?" I became very good at the perfect combination, the perfect response to make sure whoever it was got packing early. Whenever a question was posed—"What do you think about . . .?" or "How are you feeling about . . .?" or "What would you like . . .?"—I paused for a few moments, watching the asker's reaction or lack of one as they froze waiting for the answer, and then said something vague like, "What was that again?" After a while, out of exasperation and necessity, the inquisitor often just made the decision herself and went away.

The other benefits to the confinement varied. Because I was limited to my room and a curved hallway about twelve rooms long, I never felt guilty about staying in when it was sunny outside. If I needed something, there was a nurse only a button away. Guests had to come to me. I was almost always glad to see them, they rarely arrived without gifts, they always had extremely clean hands, and they never overstayed their welcome.

Because I couldn't eat or drink anything dairy or uncooked, my decisions about meals were easy. I tried not to notice that there was a vacuous rotation of meal choices that did nothing to appease my craving-based appetite. I was offered the same thing every five days, depending on whether limp boiled green beans or ashen tasteless broccoli were the vegetable of choice to accompany my over-baked chicken. On days when I was feeling okay and remotely like I wanted control over anything, I craved four-cheese pizza with extra pepperoni in every vulnerable, rapidly multiplying, and possibly still cancerous cell in my body. And on the days that I didn't want to gag from the smell of chemicals and plastic, I would've sold my soul for a big glass of Cabernet, or even a light beer. When these waves of desire and fancy rushed in, I closed my eyes and wished again for my comfortable semi-comatose state of simplicity. Most of the time, the doldrums of the ward worked well enough.

Most of my days and nights were mapped out for me. My mornings started at about 4:30 when the nurse flipped on the harsh fluorescent light for my daily blood drawing and weigh-ing—done early in the morning when human bodies are at their truest weight. The sample was sent to the lab where all my counts, including my white blood count, were determined. Following this ritual, I fell back to sleep because it had taken me until two o'clock to relax enough to sleep at all.

At about seven o'clock, my breakfast was delivered and one of the nurses talked to me about how I slept, how I was feeling, what I needed. Then, at the culmination of the visit, my white cell count was posted on the dry-erase board by the door. This was the most important part of the day.

Following the high-dose chemotherapy I completed my first week in the hospital, my white blood cell count dropped rapidly down to zero. I likened the intense chemo to pumping me full of a gallon or so of mild acid—cleaning all the gunk out of the pipes, getting rid of all the yucky germs and tumors that have built up in there. Similar to my previous chemo, it got rid of some pretty important stuff too—my hair, my stomach lining, my ovaries, and my immune system, among them.

To grow back the most important of these, my immune system, we had to plant a seed. I was given two large bags of my own bone marrow harvested before the chemo. Then, Julie, my nurse practitioner, hooked up a Texas-sized syringe to one of the tubes hanging out of my chest and slowly pushed the thick milky stuff back into my immune-deprived body. It took a couple of hours, hurt every once in a while, and my mouth tasted like fried bananas and creamed corn for a week.

After that, every day the count that told me how many white cells I'd grown was posted on the board with the rest. I watched it rise to 0.2, then 0.3. We never knew what to expect; it often rose, just as often it fell, but it always rose again. What I never forgot was that once I maintained a white count of 4.0+ (4,000 white blood cells per unit) for three or more days—a count of 4.0 or more is in the normal range—I was on my way out the door. Every day, I held my breath as the nurse posted the count, hoping I was one day closer.

Following the daily posting, I washed and got dressed. My hair started to fall out soon after the chemo, so I shaved it all off again eliminating the need for brushing, washing or bending over the sink. My wardrobe consisted of identical soft cotton leggings, shirts and socks in various colors, and some sort of stocking cap to ward off the cold, dry air circulating in my room. I typically resisted the decision required here and wore the first thing I saw.

The remainders of my mornings were filled with scheduled visits—doctors, nurse practitioners, nutritionists, and counselors. I was constantly hooked up to fluids, so rarely thirty minutes passed without having to climb over the tubes to the toilet.

Mid- to late afternoons were my only unscheduled time, the chance my mind had to work for itself. On days when I felt weak, I'd simply turn on the television and stare in a vegetable-like manner. But most days, even channel flipping, the news, or the O.J. trial didn't capture my attention, so I would draw, read, write poems, or listen to books on tape, my hospital attention span being too short for more than a couple of pages in an actual book. I'd brought about a dozen videos and kept one of the ward's VCRs in my room. People brought light-hearted movies to me as well, including a coveted copy of *Saturday Night Fever* from Zoe, my borderline crazy but very good friend from high school, several Disney classics, and a dozen brainless '80s comedies—nothing too slow paced or requiring an expanded attention span. To get my required exercise, I often walked up and down the ward, rolling my bags and tubes beside me and peeking into patients' rooms to see what their white count was.

The dark room next to mine was a source of constant curiosity—the shades always drawn and the lights always off, with everyone desperately trying to keep the glare of the outside world from reaching whatever poor soul was in there. Or maybe the tomblike container was meant to better enclose the hopeless sickness and make sure it didn't spread out to those of us with potential. I slowed during my walks long enough to peek into the room, and I watched nurses going in and out, trying to discern what they were carrying. The drawn faces of the hospital personnel weren't encouraging, but my hopes were raised when I saw a few inches of the patient. Those hopes lasted only a few seconds once I realized that I was lucky enough to see only the wrinkled and red eyes of a very thin and sickly woman. Who knows how old she was, but she looked ancient. She seemed not wise with the understanding of the cycle of life and the acceptance that it was her time to go, but miserable with the pain of her disease and the sadness of her confinement. A few days later, the room was empty, and the light from the open blinds shone all the way into my room.

Down at the end of the hall, I usually stopped in to see Carrie, a woman only a year younger than me who had been given her sister's bone marrow. Her room was bright and yellow with a balloon bouquet she'd brought with her,0 and she always had music playing and CDs strewn all over her bed and nightstand—Ben Folds Five, one of my favorites that I hadn't brought, was also one of hers. Strangely, we only ever talked about how we were doing, how long before we got out, and what our counts were up to. We never talked about anything outside of the ward, though I was constantly curious if she had a boyfriend, a dog, or brown or blond hair (hers was shaved and her eyebrows gone like mine). Even so, we mutually and silently agreed never to discuss the outside world. Our visits were usually pretty short.

Often I had visitors in the afternoon, and I was happy for the distraction at first. They asked how I was feeling, and I gave them a brief answer lacking details. There just wasn't that much to tell besides my own emotional traumas that I was avoiding thinking about and that I had learned long ago weren't what they were asking about. Usually, I resisted reciprocating the inquiry, fearing that the details and complexity of the outside world would interfere with my intentional mindlessness.

Once the conversation hit a dry spell, I would finally ask about what was going on in their lives. The visitor would breathe a sigh of relief at being able to talk about what they wanted to talk about. At first, I resisted absorbing the answer, often concentrating on the tops of their heads or staring at a scar I'd never noticed before. Inevitably, the visitor mentioned someone I knew or a restaurant I liked, and I couldn't help but listen. My ears began to perk up, and my eyes lost their glaze. The color returned to my face, and I began to sit up straight. Before long, I was listening with rapt attention and requesting details about what someone said, what their swordfish tasted like, how many people were invited to their wedding, who skipped the office party, and who was pregnant.

Then, when I was leaning forward about as far as I could go and ready to start writing a daytime drama based on the juicy details I'd just heard, the visitor would leave me. They'd leave me alone with my quiet little room and hallway and dry-erase board in my safe and dull-as-dirty-off-white-paint little world with my VCR and books on tape. They'd leave me with the news softly playing over my head and the photos of those lucky suckers roaming around outside the world staring at me from my picture frames next to my bed. Worst of all, they'd leave me with the scandalous and peculiar details of everybody else's lives swimming through my head.

Then it started. My forced mindlessness receded, and I started to miss it. I missed the outside world more than I missed four-cheese pizza. I missed it more than every short story I'd ever read and every sketch I'd ever scribbled. I missed it more than I ever wanted to listen to another book on tape or put on another comfortable pair of leggings ever again.

Almost always, Sean's face popped into my head. In one instant I missed him terribly, and in the next I resented him so completely as the representative of all those able to move beyond, and above, where I was able to go. When he thought his half-girl-friend's health crisis was over, my half-boyfriend had packed almost everything he owned into boxes in his aunt's attic, the rest into enormous duffle bags, and bought a plane ticket to New York to pursue his acting career. When we said our good-byes, I only drove away and came back once, which I considered an accomplishment, and I knew I was being selfish in doing so. Despite my tearful good-byes and secret wishes that his only dream was to stay with me rather than go to away, I knew that walking down Ninth Avenue with a backpack over one shoulder and a copy of *Backstage* clutched in his hands was where his dream took him, but part of me hated it because I wanted to be there with him. To be anywhere but where I was, but especially with him. I wished I could be a part of making both our dreams come true, and staring at the white shadowed walls of my ward made it impossible to forget that it wasn't possible. Now that he was already gone, already beyond and above, it probably would never be possible to catch up.

I wanted to have to make decisions and mistakes—get an opportunity I couldn't pass up, move to an exciting place, and follow my dreams like Sean—and experience the resulting triumphs and heartaches. It didn't seem like that much to ask, but I wanted to make my own choices and have control and not have the words "vacuous" or "unchallenging" ever enter my head again. It was excruciating.

Then, every day, it passed. I scrunched my eyes closed, and, eventually, the crushing dulled to a nice unconscious anxiety. Then, slowly, I was able to return to my surface, peaceful, mindless, simple, separatist state. Just as the quiet returned, the nurse would bring in the chicken and broccoli I ordered to seal the deal. She'd look at me strangely and then ask me if I was doing okay, and I'd pause before saying, "What was the question?" Worked every time.

Before long, it was two o'clock in the morning, Sean's photo was face down on the nightstand, and I was relaxed enough to sleep. As I'd drift off, I'd remember I had only a couple of hours until the early morning weigh-in. And only a few hours after that, the magic number would be posted prominently on the dry-erase board telling me whether or not I was one day closer to freedom. It was the best part of the day.

hi, my name is rebecca . . .

It was almost three weeks into my stay in the bone marrow transplant unit when I hit a wall. My white blood cell count was up to the consistent low 2s, with an occasional jump as high as 3.2, and I still had my eye on the door for the day I maintained that 4.0+ long enough to appease the doctor.

I was getting more and more impatient and moody, and my average attention span was down to about four minutes. That didn't help to increase my immune system enough for it to be safe to throw me to the bacteria wolves of the outside world. So, the counts just lay there like a wet, flat noodle, making me wait.

For a while, I was content enough to fall prey to the routine of the ward, read and draw, and take occasional walks down the short curved hallway, bags and wires in tow. For about a week of that time, my mother stayed with me in my room, so I had a welcome distraction.

My mother was never an overly protective or overly feminine type of mom. She wasn't content to put all her energy and intellect into her kids, living "only for us" and then wishing we'd come back when we moved out of the house. As a kindergarten teacher, she was naturally nurturing, but in a tough-love, disciplinarian sort of way. She talked to us about anything and everything, asked lots of questions about our friends and love interests, and tearfully shared almost anything about herself you wanted to know, including intimate thoughts, difficult emotions, and perhaps controversial social policy opinions.

She was a good mother, very supportive and involved in our activities and our education, but never easily fell into any of the categories of the day: Soccer Mom, Stage Mom, Stay-at-Home Mom. She always tested the conventions and us and placed more emphasis on intellect and individualism than on appearance or social skills. While I had no guidance on how to put together a smart wardrobe or apply lipstick, from an early age, I was thoroughly prepared to confidently tackle the academic world.

Intellectually, I felt very close to both my parents, but hugs were saved for long hellos and good-byes. "I love you"s were only shared in birthday cards and at the end of the occasional lengthy long-distance phone conversation. It's not that I didn't know my parents loved me. They just showed it in different ways, saving physical and verbal affection and gushy compliments for when they really meant something.

But they were always quietly and stoically present. They were there and not there when needed during the first rounds of chemo and radiation—always watchful and worried, but never showing too much concern in my presence. They pushed the "reserved" envelope a bit when their child had a recurrence and had to be put in the hospital for a long period of time. They decided to trade off staying with me throughout most of my stay; and there was no question about it.

When they told me this, they said that their practical nature reasoned that I'd need someone eventually when I started to get weak from everything happening in my body and before the replaced bone marrow started to make me stronger. Of course, they were right and I agreed. I didn't know to what extent this caregiving would go.

The night after the official bone marrow transplant—remember the Texas-sized syringe filled with thick milky stuff?—I spent an entire evening and into the night bent over a bedpan heaving a steady stream of thick milky stuff and what the nurses told me were probably pieces of my stomach lining that had been killed by the early days of high-dose chemo. My father had been there the entire day of the transplant, watching my face as I winced during an instant of discomfort and listening to me when I said I was okay and he could leave. This was before the heaving. After a couple of hours of regurgitating what could have been pieces of what I considered an important organ, I called home. I didn't have an agenda. I was just scared and unhappy to be alone for the first time since I'd entered the ward, and probably crying. My father answered the phone. I never asked him to, but he immediately got into his car and drove the fifty miles to Cincinnati to sit with me. There wasn't one damn thing he could do, and he knew it, but he knew we'd both feel better if he was there. He was right. Only after the heaving stopped late at night, and only after I fell asleep from exhaustion, did he drive the fifty miles back home.

A couple of days into my mother's stay, I became pretty weak. For a little over a week, I found I needed someone to hold me up while I took a sponge bath, and I got breathless hopping off the bed. I needed my mother there to help me, but more than that she was a welcome comfort and distraction. But I didn't tell her that.

While she was still there, I began feeling the sores. Luckily, none were visible, but they felt like my plastic cafeteria chicken carried a machete with it one day and, on its trip down, swung freely at anything pink and vulnerable. If indeed all things are relative, then it could have been much worse. I'd received warnings of sores on my lips, mouth, throat, esophagus, bladder, urinary tract—anywhere and anything vulnerable. I suppose I was lucky that it felt like steel frets were driven into *only* my throat and esophagus. When my fourth milkshake of the day stung like witch hazel on its way down, I stopped eating altogether.

It was one of my favorite nurses who said one day, "You know, we can put you on a continuous morphine drip for the pain. . . ." Continuous? Morphine? Having not had a drink or an orgasm in several weeks, I figured drugs would be a nice distraction. Besides ridding me of the pain, I guessed it would rid me of the boredom to an extent as well, or at least make me not care if I was bored or not. And I knew my mother would soon be gone, and I wasn't sure how I'd react. I quickly agreed to The Drip.

I'm not a big believer in drugs. I smoked marijuana twice during college, and only when the beer at the party was gross and cheap, and I was desperate to avoid a hangover because I had to write a paper the next day—I was a party animal. But that morphine was some great stuff. The drip started off at Level One, then I convinced them to up it to Level Two, and, only half-legitimately, upped it again to Level Three. That's the highest I got, but "high" is a great word for it.

Level Three was very, very comfortable. I've never been so blindly optimistic and relaxed in my entire life. I smiled a goofy grin and offered a "good morning" to my baggy-eyed nurse during my daily four o'clock weigh-ins. I laughed at the O.J. trial, my pie-in-the-sky attitude convincing me it was more like a movie than real life—"No one could do anything that horrible." My jaw hung open as I napped happily throughout the day. I caught myself dozing during phone conversations and visits from my nurse practitioner and then not really caring about the gaffe when I woke up to someone's annoyed barking because I'd fallen asleep on them.

Sean traveled from New York to visit me for a few days, sleeping in the same cot my mother had, and I smiled and laughed and caught myself snoring during catnaps the entire time. I still don't remember too much about his visit, other than it made me very happy to see him, and I babbled about it and our "perfectly wonderful" half-relationship with one of the younger nurses the day he left. She agreed he was a great guy for coming all this way on a starving actor's living, not to mention extremely handsome, as she looked at his headshot propped on the nightstand. But, after hearing our relationship history and then listening to my morphine-induced optimistic realization that he absolutely was going to buckle down and make a commitment to me someday, she only looked at me skeptically. I didn't care and regularly leaned over and kissed his headshot.

I didn't draw nearly as much, but when I did, the angry dark lines disappeared and became shining suns and smiling, healthy faces. Even the flesh-rippingly condescending nurse I disliked was suddenly very dear to me, and I told her so. At night I dreamt of walking through grass with my shoes off, eating fruit straight from the trees. One dreamy night I stood at the top of Notre Dame, and my hair, at least seven feet long and wavy, blew gracefully around me as I smiled into the wind and looked down the Seine and ate slice after slice of rich French cheese. As I descended the stony steps, a shirtless man scooped me up and walked me to his palatial room in the Latin Quarter. Every moment was as real to me as the really long movie on the television where the former football player from *The Naked Gun* movies is accused of murders he couldn't possibly have committed.

My quarantined world became a bright green and gold utopia where everyone, including myself, was always going to be okay. After a while, I began to wonder what all the fuss was about and considered just walking out.

When the nurse came in to tell me they were going to phase me off The Drip, I actually surmised how I might get a solid enough grasp on the phone to swing it across the bed, knock her out, and tie her up. "Really, my throat is still so very sore," I pleaded, whining with fake tears in my eyes, but she wasn't buying it as I wiped away the drool that accumulated during my nap. The other nurses hovered outside my room, and I knew they'd discussed this as a group—a conspiracy. I whined and pleaded, but finally was forced to quit cold turkey.

From being on one of the lowest doses of morphine available in the unit, I suddenly felt like a junkie drying out in rehab. Within an hour I was crying, wondering why O.J.'s prosecutors were so frustrated—any jury would be able to see he's a guileful murderer. "God, why can't they see it!" I even screamed at one point, causing the nurse to come in to see if I was all right. I stopped answering the phone and complained with every nurse visit that my throat was killing me. Couldn't they see me shaking? Come on, just a Level One, just to take the edge off. . . .

The second night I was off The Drip, one of the television networks aired a special anniversary showing of *The Sound of Music*, including out-takes and interviews with cast members. While I adore musicals, I somehow had acquired disdain for this one. But it was undoubtedly a favorite in my family—right up there with *The Wizard of Oz* as a popcorn-popping, sitting-together-around-the-fireplace classic—and I was a goner. From the minute Julie Andrews showed her virginal face on that Austrian hillside until the last credit rolled over photos of the actual von Trapp family following their escape and immigration to America, I blubbered like a baby. It didn't stop for two days.

Nurses came and sat with me, including the patronizing one I claimed, under the influence, I adored. Family and friends began phoning repeatedly when no one answered, and I never returned their calls. The woman who cleaned my room every afternoon patted me on the hand and put a new box of tissues on my nightstand.

They all believed I was heartbroken because my counts were on a plateau. I'm not sure they ever realized that I just wanted a hit—of what I wasn't sure. I desperately wanted to escape reality, and no one was letting me. Not to mention that it seemed almost cruelly appropriate that the only satisfying nonalcoholic synthetic high I'd ever experienced just happened to be during the most difficult few weeks of my life. What was the point in going back to reality at that point? It took drugs and an Austrian romp on a hillside to show it to me, but my current reality really sucked and neither the drugs, my parents, nor Sean were around anymore to dull the pain.

Every addict, even if self-appointed, has to hit bottom at some point. I think that was my worst moment—the moment I realized it was just me that was going to get me through whatever lay ahead, and I didn't like that. Yes, I'd gone to my House of Prayer in the woods and summoned my internal forces, gathered my strength and courage to face the possibility of pain and death. In a big-picture way, if the hand of death ever came through my door, I was golden. Day after monotonous day, as I tried to get through the morning and then the evening and then the dark and quiet of the night without losing my mind from anxiety or the frustration of not knowing if or when I would get out, I wasn't.

I finally stopped crying. I finally returned to my state of conjured peace and minute-by-minute summoned strength. I looked at the photo of my outdoor cathedral and pretended I could hear the spring breeze whispering over my head and smell the mossy earth beneath me. Okay, already, I did it. And it worked. But the easier way always was only a nurse or two away, and I thought about it every day. I never stopped craving The Drip.

chapter eighteen

sitting beside me

My cat loved my sister more more than me. I couldn't blame him.

My sister is four years older than me. One of my earliest memories is lying on her lap on the sofa and my parents taking a picture of us. I don't remember her holding me that often when I was little, but I do remember sitting beside her for most of my childhood—the problem with having only one sibling. You're forced together as the only nonadults in every situation. In the sticky backseat of the car during autumn Saturday road trips to apple orchards, on every amusement park ride I ever rode, on the living room floor watching *The Muppet Show*, picking strawberries in the garden in the hot July sun, at dinner, kicking each other underneath the table when it was the other's turn to say grace. And, despite the age difference, we still played together a lot.

Neither of us ever was very girly, never owning any Barbie dolls or fingernail polish. We preferred bikes and kickball and playing "guns," which was really just multiple reenactments of *Star*

Wars, with the boys in the neighborhood. Lin had an overstuffed doll named Baby Sue that was beaten up and nicked from all of its abuse, and I had an enormous collection of stuffed animals, mostly centered around sea creatures and teddy bears, but that was the closest we got to traditional girly play.

We doted over our dogs—a Dalmatian named Mandy when were kids, and then an eccentric little beagle named Val acquired less than a month after Mandy died—and played with them together as much as they could take, frequently competing over their attention.

We both liked books and music, preferring when the two were combined in an album and a big book to follow along, like *Free to Be You and Me, Multiplication Rock,* and the Walt Disney movies of the time. My favorite was *Robin Hood,* but Lin liked *101 Dalmatians* because of our own. Cruella DeVille didn't give Lin bad dreams, as she did me.

This obsession with the spoken word got larger as we got older, and we each received our own tape recorders one Christmas. At first we used them on our own. I don't know what Lin did, but I have at least one cassette tape filled with me reading my favorite Dr. Seuss books—*Fox in Socks* being the most challenging, and I tried to read it faster and faster without fumbling—and *James and the Giant Peach.* Then one day we decided to pool resources. We spent hours pretending to be reporters and interviewing people, including everyone from school cafeteria workers to Barry Gibb. Baby Sue became a central character in spontaneously created soap operas and Brady Bunch–like sitcoms, and friends who came over would get into the act. Somewhere there is tape after tape of this stuff, and I have no idea where. My worst fear is that I will spot them at some garage sale in the San Fernando Valley.

Then, my sister became a teenager. The reenactments and Baby Sue Olympics ended. She didn't just become a teenager bored with her little sister and with breasts too big to run fast. It was much more complicated than that. One evening during her thirteenth year, my sister rolled over as she was lying on my parents' bed talking with my mother and began to shake uncontrollably. I only saw the end of the episode since I already was in bed, but by the time the emergency crew arrived, it was over. They said she'd had a seizure, but didn't know why. After some tests in the hospital, a doctor told us she had epilepsy, and what we'd seen was a grand mal seizure. He comforted us by saying they wouldn't all be like that. Smaller seizures, called petit mal, would only make her stare strangely and maybe subtly turn her head one way or the other. Usually, epilepsy is apparent from birth and is grown out of by adolescence, he also said. Since the seizures didn't start until adolescence, she might never grow out of them and probably would have to take pills for the rest of her life to control them. My sister didn't take this news very well.

My parents urged her to start practicing taking them several times a day by swallowing pill-sized pieces of candy, and she only complained a little. They started watching her more closely and checking in with her more often when she was in her room alone for a long time. Usually, she just had petit mal seizures and snapped out of them after a minute or two with a horrific headache. Still, we didn't know exactly what to do. As a family, we were nervous and awkward about how to handle it in public, usually trying to stay calm and get Lin to a safe place as quickly as possible. Once, she had a petit mal during one of my band concerts, and Dad just put his arm around her and put her head down on his shoulder so she wouldn't be so confused when she woke up. They went home quickly after the concert.

No one could tell us what set the seizures off, or even if there was any external trigger at all. We just went along our merry way, trying not to make a fuss and not talk about it much. Normal is good, right?

Who knows if this was the right way to go about it. Life with epilepsy isn't the end of the world, but it might be to a thirteen-year-old who's just heard she has an incurable and only somewhat controllable brain disease that makes you shake violently when you least expect it. By definition, teenagers have an overly dramatic reaction to the world and take themselves way too seriously. Lin was no different. During junior high and high school, when most people are dating and creating lifelong friends, my sister retreated inside herself—spending hours studying, coloring in her anatomy book, and reading murder mysteries. She had a few friends, even fewer boyfriends, and never wanted to play anymore.

I was no different as a teenager as far as the taking-myself-seriously goes, but I didn't have such problems to consider. I made friends with Zoe, one of the most outgoing and controversial figures in the performing arts program that shaped my social life at the time. While I was nearly always in her shadow, I took her lead and decided to pretend to not be shy. It's easy for a closet introvert to be unconventional or outrageous when you know someone else will trump you every time. In the days of pegged jeans, Zoe wore her wide-legged pants rolled way up her legs to make them shorter and show off her shapely calves. I came to school some days wearing two socks of different colors—both matching my outfit, of course. Zoe wore sexy hot pink lipstick to match her hot pink sweatshirt with giant lips on it, so I wore green eyeliner to match my green striped rugby shirt. When Zoe became the lead in the school musical and I was slated to be first chair clarinet in the orchestra, I insisted on auditioning anyway and taking a small role just to be on stage.

As opposed to seeking the safety of the introversion that came naturally to me, I pretended I was an extrovert so that no one would know I wasn't ready to be one yet. Besides, we already had a family wallflower.

By the time Lin left for college, I'd lost count of how many times she'd attempted suicide. She was never very good at it by design—swallowing an entire bottle of her epilepsy medication less than an hour before she knew someone was due home, jumping from first-story windows, hoping the knife she was holding would happen to nick a main artery, scraping up her wrists and then walking into the bathroom while I was peeing. But no matter what we or numerous therapists said, she had convinced herself that dramatic, self-destructive actions such as these were the only way to get the help she so desperately needed and still feel good about asking for it.

As she got older, the health problems continued to increase, but she'd stopped the attempts. By the time I got lymphoma, her epilepsy was largely controlled with little or no medication. But she'd added to her list of diagnoses chronic depression, ulcers, migraines, a thyroid imbalance, various female problems, and lactose intolerance. The emotional cage she'd built for herself, understandably, was the most affecting. But we didn't realize to what extent she'd caged herself until one day only a few months before I was diagnosed. That was the day she called my parents, one week before she was to marry a very unusual but seemingly well-intentioned man, and came out of the closet by announcing she was in love with the woman who lived next door with her estranged husband and two daughters. My mother and I helped her call the guest list and return all of the gifts.

As I got older, I became more comfortable with my closet introversion. The type-A part calmed down a bit—but only a bit. Lin created a very busy and people-filled life for herself. Nevertheless, our family roles remained the same. I was the daughter Mom and Dad used to tell, "We never worry about you"; the one working at a competitive job I hated and saving myself for a man with a fear of commitment greater than his fear of me dying. They never said they never worried about Lin, but she was the one in a loving relationship who had found a good job working with residents of a nursing home—"taking care of people who need help more than I do," she said. I think complying with our roles was easier than getting to know each other.

By the time I got sick, we fit at different ends of the dinner table, balancing it with opposition and awkwardness. And we rarely ever spoke.

When I moved home with my parents, we still weren't close. After years of living with my parents, Lin eventually had made a home for herself and her new family—her new partner Tami and her two young daughters. She wasn't a big part of taking care of me, not that there was that much to do besides make sure I had a place to bathe and sleep where I didn't run into too many bacteria, but she was always on the periphery. She lived close by, so throughout those months she occasionally helped with errands and taking care of the house or dog if I was in the hospital or my parents had to care for me somewhere other than home. She and her family were definitely always around, and I appreciated it.

The biggest help my sister offered was taking in my beloved cat Lucifer—a fairly large beast not welcome in my parents' house by order of my father and our elderly family dog. My sister's family had a few cats, and Lucifer, a beautiful and proudly strutting tomcat used to being the master of the household, was an interesting new player. He quickly took his place as the alpha male and strutted through the wide rooms and over the bookcases with his bushy tail sticking straight up in the air to demonstrate his superiority. Their house was like the African plains, and Cifer was the man-lion who lay around digesting all day, impressing everyone with his beauty.

I visited briefly every couple of weeks to check on my beloved kitty and to be sure he hadn't forgotten about me. I'd walk in the front door and, barely able to contain myself, wail

"Kittykittykitty!" at the top of my lungs. Instantaneously, Cifer would appear from nowhere and run toward me, meowing tentatively, and then pounce with all his might from the floor to my shoulder, where he knew I'd catch him and hold him there like a toddler so he could purr in my ear and rub his cheek against mine. He'd slept next to me on the sofa when I was too weak from chemo to get up to go to bed and licked my bald head to wake me up to feed him. He knew I needed him. Every day before I was diagnosed, he tried to sneak outside when I came home. Then, the day after I was diagnosed, he stopped. He was my baby and, with Charlie the ailing seven-year-old goldfish having been mercifully flushed during a mood swing, the only house companion I still had around who'd seen me without hair. Sometimes he was the only reason I ventured to my sister's house.

Only a couple of weeks after getting out of the bone marrow transplant ward, I made a trip to my sister's to visit the family and Cifer. I hadn't seen any of them since before I'd gone into the hospital and was secretly afraid that my cat had forgotten me. It had been longer than two months.

When I walked in, I hugged my sister and Tami, and, not able to wait, called to Cifer. "Kittykittykitty!" He didn't come running. I called again at top voice. "KITTYKITTYKITTY!" Nothing. I started looking around, under the sofa, behind the closet door. Nowhere to be found. I looked at my sister accusingly and asked where he was. She looked back without a lot of surprise, but a little defensively said, "I don't know. He's here somewhere." After a couple more minutes, my sister sighed and took a breath. Then she let out a call I didn't recognize and a version of his name I didn't like, "Louielouie!"

Only a moment passed before Lucifer came out with his signature stealth prance, and I readied myself for the weight on my shoulder. But as he approached, I realized my worst fears had come true. He jumped onto my sister's shoulder.

I tried hard to hide the fact that my eyes were welling up and my jaw was clenched with competitive anger as my sister walked around with him purring into her ear, but I started to lose it when Cifer sniffed at me like a stranger when I tried to pet the bushy fur on the side of his face. Tami said that she was sure if I stayed long enough he'd warm up to me again. It's just been a while, they said. And I did look a lot different than the last time I'd seen him, Lin said. I probably even smelled faintly of strange chemicals still—I know I could still smell them. I'd also lost about fifteen pounds and was still hairless. I'd worn a big sweatshirt and a baseball cap and was still shivering a bit.

So, I sat down. My eyes never left my cat, now walking around the room not noticing me, and I actually talked with Lin.

I told her all about the hospital, the doctors, the never-ending vitals-checking and procedures. I described all of these things as if I were talking to one of my friends who didn't go into hospitals and whose most serious illness was the flu. She quickly corrected my patronizing demeanor and word choices, however, by nodding knowingly. She'd been through a good many needle sticks and specialist visits in her day. I told her about the nurse who whispered to me about the pills she was giving me as if I had Alzheimer's disease and didn't remember the previous nineteen times she'd given me the same medication, and the hospital-appointed psychiatrist who sent me bills for two hundred dollars every time she popped her head in to say hello even though I told her I didn't feel like talking. I realized after I'd gotten out that my insurance wouldn't cover in-hospital counseling, though that wasn't the reason for the refusal.

"Damn insurance," Lin said, and I nodded, but returned the complaint with, "Damn psychiatrist." To this she gave me an enthusiastic nod that I thought might give her whiplash. I didn't hold as serious a grudge toward health professionals of any sort as she did, but I knew she understood the pain and embarrassment involved in baring one's body and soul to strangers when most of your life has been a carefully orchestrated attempt to keep such things to yourself. Taking that risk had cured me. But Lin had done most of her curing on her own and never seemed to find a counselor she didn't view as a threat. Mine was a solitary and singular health problem only a little over a year old. Her problems were numerous, nebulous, and interconnected, and had debilitated her to some extent for most of her life.

I looked at my sister as she talked and for the first time noticed her jaw, not square like a soap opera actress, but strong and set with determination. Her blue eyes and dark eyebrows—borrowed from Grandma Kirby like mine—were furrowed with what was either concern or ferocity. She wanted something—health? privacy? a good life despite obstacles?—so badly and seemed singular in her purpose to get it. I'd seen that same face in the mirror more than once.

While I was in her home, Lin brought me food, juice, and a blanket. She loaned me a couple of books and videos and gave me some homemade cobbler wrapped in foil. I sat on her sofa and talked with her, let her bring me things whether I needed them or not, and waited for my estranged cat to come by for a visit. He never did.

Then, as I got up to leave, Lucifer finally sauntered over and purred around my shoes. I'm not quite sure if he recognized me or if I was just another warm body to pay attention to the alpha male of the household, but I couldn't contain my pleasure. I picked him up and nuzzled in his fluffiness before thanking Lin and Tami for everything and walking out.

When Lucifer moved back in with me in Cincinnati a few months later, Lin and Tami only visited him once. He was happy to see them, even though Lin called him Louie. He purred and meowed, and I think I saw Lin's eyes mist up a little. I only misted up when she said good-bye and that she loved me. It was the first time in years she said it without the catalyst of a crisis to make it seem appropriate. I happily said I loved her too and watched them drive away.

My sister and I still don't speak that often, but I'm glad she's there to sit beside me when we do.

chapter nineteen

back to normal, already?

On April 26, 1995, my doctor announced that we had beaten the non-Hodgkin's lymphoma I'd fought for a year and a half. There was no fanfare, except in my imagination. The grand cathedral bell didn't toll. No tickertape fluttered from above, and no rose petals were thrown at my feet by wreath-adorned children. At home, no sequin-clad drag queens chanted my name and handed me a frothy drink while coaxing me onto the dance floor.

I simply hugged and thanked my bone marrow transplant doctor and the nurse practitioner with whom I'd spent much time during my eight weeks in the hospital and left. I'd won the fight against the cancer, at least for the time being. Time to get back to normal, right?

Normal is still relative.

Only two months later I found myself back at work, commuting an hour each way from my parents' house for two or three days a week. The motivation was not my own yearning to return, unfortunately. When I started my leave of absence in February, I knew I had to be working full-time—at least thirty hours a week—by July or I'd lose my health benefits. I started working my way back to full-time as quickly as I could. But with reality being too business-cold, I justified it in about a thousand ways—regaining my independence, putting the past in the past, recovery through distraction. My family and friends helped. Joanie and Fiona smiled and said quickly, before I had a chance to doubt the idea, "It's a great idea. Get yourself back in the world." My parents weren't thrilled with the commute or the forced nature of the return, but I don't think they minded being rid of me for a couple of days a week.

The biggest reason I convinced myself it was a good idea to go back to work so quickly was that I'd be gaining control of my life again and really living. Technically, I was living already. This was something that never quite felt as novel as a miracle to me, but that everyone wanted to label as one. On those days when I understood this notion, I felt enormously fortunate. But even on a technicality, living in a room in your parents' house with no way to support yourself, no actions outside the watchful eye of your concerned Mommy and Daddy, and practically nothing of your own nearby, is not really living one's own life.

So, off to work I went, as spruced up as I could be. I'd stopped wearing head coverings a couple of weeks before, even though all I had on top was enough to claim that, technically, I wasn't bald anymore. They'd taken out the triple lumen catheter so there were fewer wires and gadgets to hide underneath my clothing. I went shopping for some new business clothing. Got the car tuned up. I was as ready as I could be to return to life. Whose life it was, I wasn't sure.

My first day back on the job was okay—not good, not bad. No fanfare, which was more than fine. There were a few streamers and signs on my desk welcoming me back, which I was grateful for, but I was eager to find out what had happened while I was gone, who everyone was four months later, where my new immune system fit in now, and why I'd chosen to go back to a place that made me unhappy even before I was sick. I attended meetings, talked to people, signed some papers, and started to make it official. Things had changed—several new people had been hired, a dress code scandal had occurred and made it okay for women to wear pants, my pro bono project had been a miserable failure, and I think I heard a voice I didn't recognize ask who the bald chick was when I walked through the office. At lunch, I went to my car, practically bouncing because I knew people would be watching through the floor-to-ceiling glass, and put my head down on the cool steering wheel for what I thought would be just a quick break. I woke up forty minutes later and knew I could have slept another hour. Other than that, things went fine. Everyone was kind, and I ended the day exhausted but strangely satisfied with the results. One step at a time.

I was on my way out the door when Josh, an intellectually oriented forty-something member of the agency and one of the people who hired me a couple of years before, stopped me and asked how I was. I breathed a sigh of relief to see a familiar face among the newness and started to tell him how exhausted but strangely satisfied I was that I got through an entire workday without keeling over. He looked me over as I talked, and I stopped because I could tell he wasn't listening to me. His eyes finally stopped on my head and he asked, "So, when did you stop wearing your wig?" I laughed and said the last time I'd worn it was my last day of work in this office. He smiled and said, "Well, I'm glad. You seem more normal now. It's good to see the real you around the office again." My next day at work, two other colleagues offered almost identical welcomes. One of them even said, "Now you're normal again. You can finally get back to business." I wanted her to ask next how exactly I'd been so horribly abnormal before, or if I was even ready for "back to business," but I didn't.

That weekend I called Sean to report on my first couple of days back on the job. His first words were "How does it feel to be back to normal again?" I chose not to acknowledge those words and went directly to Josh's comment. Before I could express my frustration, he said, "God, I know what he means. You're looking great and feeling great. Thank god we can all put all of this behind us." I had been cancer-free less than two months, he hadn't seen me since I was in the hospital, and I'd never once said I was feeling great. Not once.

Clearly, I was up against something bigger than myself. What were my post-cancer expectations? I wanted to live three months at a time without hearing the words, "It's back." Beyond that, I wasn't quite sure yet. But everyone else's needs were more complex. They had a quest, a burning need it seemed, for me to get back to "normal"—looking, acting, talking, functioning just like I did before all of this happened. The quicker it happened, the quicker they could forget about everything. And they apparently expected me to do the same. I understand the impulse, believe me. Until these expectations were forced upon me, I thought I also wanted it that badly.

After the success of the transplant, my "being just so darned happy I was alive" elation lasted only a few weeks. But the novelty of survival disappeared exponentially over time once the snags of just living—like the possibility of losing my health insurance—became more regular.

But it also was deeper than just "snags." Cancer was the ultimate exercise in self-awareness, but hardly a realization that I was a saint. Turns out, when you spend a couple of years focusing almost exclusively on living to your next treatment, the potential to uncover, exacerbate, or ignore any insecurities and relationship problems is enormous. Stuck in a dead-end relationship cycle?

Whoa, can't tackle that just now, you've got bigger fish to fry. Not happy with your current career choices? Whoops, there it is, but you can't deal with that now because you're sick! When I made it out the other side of the immediate crisis, it didn't make the problems go away or become less important. Instead, it left me to get back to the problems I had before I was sick, and on top of that, to deal with everything I'd done and felt while I was sick—if this was possible at all.

Obviously, the easiest thing for everyone would have been for me to return to exactly where I was before I heard the words "You have cancer." More than anything, I wanted that for my loved ones who were waiting with bated breath to get back to a time when they didn't fill with dread when they heard my voice on the other end of the telephone. I wished with all my might that some hard-won enlightenment and sense of peace would spurn me on to a normal and more fulfilling life in no time. But I knew I was nowhere close to any of these things. As much as I tried to get them, the cancer would not grant them to me. With every passing day I felt more and more abnormal.

I started not sleeping well, and when I did sleep, I had horrible nightmares about NASA loading cancer patients hooked to strings of IVs onto rockets and shooting them into space without a destination, and dreams about riding roller coasters loaded with bloody bunny rabbits and set on fire for added thrill. I began jumping every time the phone rang and washing my hands maniacally whenever I touched a public handrail or doorknob. The week before every three-month CT scan I became wholly preoccupied with not knowing what the next year of my life would be like and spent hours plotting how I would go into hiding and relieve everyone from the burden I'd become if the news wasn't good. I returned to my support group, arranging to work on Tuesdays so I could attend the Cincinnati-based meetings after work, but spent each session completely silent because I couldn't bring myself to ask a room full of people who desperately want-ed to get to where I was to explain why I wasn't happy about being alive. And I found myself missing Ginnie. When out with Joanie and Fiona, I often was moody and quiet for no apparent reason, and they often were frustrated and annoyed in response. Phone calls with Sean became increasingly exasperating as he continued on his New York–inspired dream-catching expedition, refusing to understand why I wasn't just so happy about how everything had turned out. My parents, I suspect, had no idea what I was going through, and I wanted it that way.

Finally, I had to pull out the big guns, so I went back to what I knew—nature. Okay, I'll admit it; I went back for a quick fix, to make myself into that happy, well person we all wanted to see so desperately. It was the simplest solution. Just a few breaths of sum-mer air and a tearful chat with myself would do the trick. I went back to the nature preserve, now dripping with green as I'd hoped. Walked right back to the same spot I'd started in the day before my transplant, the entrance to my outdoor cathedral. I closed my eyes and waited for the breeze-induced calm to sweep in from behind me on cue. And waited some more. Waited. Waited for five minutes, and not one inch of air moved around me. Suddenly, it felt hot and humid, and something stank like rotten fish. I looked over to the banks of the creek to see a dead frog lying with its belly face up to me, a little bit of white-chocolate-like ooze coming out of its severed back leg. Then a large group of hikers led by a noisy little man pointing out the unusually low water level came traipsing by, so I turned on my heels and started walking back up the path to my car.

The path was short, but almost entirely uphill. My frustration forced me onward, but my heart was pounding, and my legs were weakened from lack of exercise. My whole body, now fif-teen pounds lighter than during that last trek up the hill, seemed as heavy as the water starting to fall from my eyes. My legs couldn't carry me fast enough. By the time I reached the top of the hill, I was huffing and puffing and red with fury. I think I said out loud, "It'll never be over."

Then I sat in my car and waited for more tears to come. They didn't. I felt compelled to cry more often than I liked, but I'd refused myself the luxury. It's confusing for most people if you burst into rage-filled tears when they congratulate you for surviving cancer. And I didn't want anyone to know I wasn't simply elated every minute of the day because I was alive.

Finally, I heeded urgings from Joanie and Fiona. One day Joanie said quickly, as if she'd been waiting for weeks to say it and finally found the opportunity, "Maybe you should go see someone who knows how to deal with you." Tired from trying to explain myself, I gave in. I got a name from Vic, my support group leader. This therapist had seen other cancer patients and, he added while looking at me over his glasses, handled people with varying degrees of Post Traumatic Stress Disorder. I'd never mentioned PTSD to him. I started therapy the next week.

For the first two visits I cried, finally, but uncontrollably. Through tears and stuffy noses, I tried to explain to her that I couldn't justify to anyone that I wasn't happy; that I didn't want to go to work, and woke up heavy with dread every morning of my commute for being forced to go; that my friends couldn't stand me anymore because I didn't come to terms with an eighteen-month ordeal in a couple of weeks of anxious reflection; that I believed Sean's disgust with what I'd become was so strong that he went all the way to New York, and I wasn't sure I blamed him; that I was disappointed because I believed I didn't have what normal cancer survivors describe—a life-lifting experience that propelled me to do something really interesting and affecting like becoming an oncology nurse or making a made-for-TV drama. She giggled a little at that one, knowing me well enough already to know I wouldn't have the patience to be a nurse, and the networks would never buy anything containing bloody bunnies on a flaming roller coaster. Then she handed me a box of tissues.

At the end of the second session filled with the same words as the first, she sat quietly for a moment, waiting for me to calm down. She didn't say a lot, but I could tell she had something important to share.

"Why do you want to be normal so badly?" she said.

"I don't. They do," I said. I could tell she wasn't completely buying my line that the only pressure toward normalcy was coming from employers and friends, but moved on anyway.

"You're not ready, sweetie. Not even close. And now you've managed to push yourself further away from it than you were to begin with."

It only took a second for the words to hit me, and then I cried inconsolably for the last five minutes of the session. I cried from relief.

I continued to go to my therapist once a week for a while. On the day that I happened to say, in a conversation that had nothing to do with cancer, that there was "no such thing as normal," she told me we could finally start my therapy.

chapter twenty

casualties of war

In any war, there have to be casualties.

In my own mêlée, I lost only two. But that was enough. To tell you about them, I have to go back a little bit.

Before going into the hospital, and on posttransplant days when I wasn't commuting to work, I spent hours at my own workstation in my room. My life overall was relatively solitary. I appeared to be the perfect unobtrusive sick person—like Emily Dickinson if she smiled every once in a while—and my room was a wonderful refuge from the reality of my situation. I could sit for hours and hours undisturbed, with only my thoughts to keep me company. On most days, it was absolutely the dullest place on Earth. On most days, I became agitated and nervous by midday, and actually caught myself pacing through the house pretending to look for the dog more than once.

Not too far into my self-imposed quiet, I admitted I needed help passing the time without going crazy.

Joanie and Fiona both were living and working in Dayton, Fiona, as an administrative assistant for a nonprofit organization, and Joanie, as a stage manager for a local dance company. They were busy living their own lives, and they had their own close friendships to nurture. When I'd announced I was moving home, they both were elated, exclaiming that they could "better take care of me" from so close. Now they seemed even more eager to help me preoccupy myself in my claustrophobic state. I was more than eager to take them up on it.

Even before the transplant, they proved to be compassionate friends with a welcome knack of surprising me with their blunt honesty. The bullshit explanations and acceptable responses I'd learned and practiced since my diagnosis didn't work on them, and they called me on it.

After the Evil Sitcom blowout with my father during those first few weeks of living at home, I flew into Fiona's apartment, spouting out the "horrible" things my father had said and how "perfectly unreasonable" he was being. She sat calmly waiting for me to finish, and then she spoke. She didn't let me get away with my tirade any more.

"Your house is like a powder keg," she said. She'd been there many times and knew of which she spoke. "You've got it so good over there. But you're so pissed off that you have to be there, and they're so pissed off that you don't want to be there, that you're being ridiculous. Go home and apologize to your father." I was so surprised and impressed with her instant and complete capsulation of the formerly veiled bottom line that I started to laugh. I still was pissed off at her for calling me on something I just wanted to be comforted for, but I hugged her nonetheless.

Fiona, a large woman with a sarcastic nugget for every situation and a short patience for bullshit, gladly and tightly hugged me back. We weren't friends in high school, but I remember that she always armed herself with smaller friends who weren't as aggressive or gregarious as she was and was always waiting when some social or emotional catastrophe occurred. Maybe I was her latest charge—the friend who needed her the most just then, so therefore the one who got the most attention. Whatever the reason, I enjoyed the security of her friendship.

As time went on, the three of us saw more and more of each other, especially Fiona and me. We even got into a routine. I frequently surprised her at her office during the week, and when Joanie wasn't on tour, we'd surprise her at the studio, and we'd go to lunch. We watched *The Real World* on Wednesday nights with bowls of popcorn and, upon Fiona's insistence mostly, salads. Sometimes we stayed up late playing Sega just so we could go to Denny's during the wee hours, like high school, and look for the "alternative" people we still knew and used to be.

During the day, we phoned and bitched and gossiped and worried for each other. Joanie was planning her wedding and helping to take care of sick relatives. Fiona had boy problems and family problems, so we always had plenty to talk about. Like Joanie, Fiona always had a raw and seemingly confident opinion about everything from pop music to her brother's girlfriend to Social Security. She was challenging, consistently entertaining, and listened to anything I had to say about anything. And the fact that I had cancer scared her to death.

One evening, a few weeks before the transplant, I lay on Fiona's sofa with ice packs strapped to my butt after a bone marrow extraction. She brought me chips and cheese dip to make me feel better and then sat by my feet and, after giving me a concerned look, reluctantly asked what other things I would have to do as part of the transplant. She listened quietly to my sterile and rehearsed explanation of where I'd be, what they'd give me, what my chances were, how long I'd be gone. She didn't say anything at first. Then she just said she had a feeling I was going to be okay. But the look in her eyes was one of uncharacteristic uncertainty. It scared the hell out of me.

Despite the looming reason why we were all spending so much more time with each other, I grew to depend on both of them and them on me. I happily contracted diarrhea of the mouth when I was with them, especially Fiona, and revealed more about what I was really feeling and thinking about everything but the cancer—resentment, boredom, paranoia—to her than to anyone. I also acted up on them more often than I probably should have. Like the time I flushed my seven-year-old dime store goldfish Charlie down the toilet because he was starting to float strangely and had stopped eating. He was actually starting to revive a little when I decided to "put him out of his misery." Then I cried like a baby and called Fiona to tell her what I'd done. Her reaction was surprisingly supportive considering I'd just murdered the pet I'd taken care of longer than anything, even my cat. But I knew she was beginning to imagine what I'd look like surrounded by white padded squares.

One day I misguidedly revealed my doubts about Joanie's upcoming marriage and my concerns for her long-term happiness. This was after I'd agreed to stand up with her at her wedding and only a few weeks before the event. Fiona had expressed her own doubts in the past, similarly out of concern for Joanie, but it was apparent these were unpopular opinions now that the wedding loomed closer. My diarrhea-of-the-mouth-inspired timing was misguided, but my concerns real. Fiona and I didn't talk about the wedding after that.

Because of all of the time spent together and the blunt words we all trusted each other with, I believed my friends would never judge me for these thoughts or actions. I believed they would understand that troubled place from which most of it came—that so many things being "all about me" was a situation shoved on me against my will and that I only enjoyed it every once in a while. I believed they were the only ones who would understand these things. Consequently, I shut out a lot of other people.

Throughout those months of quiet and recovery, I received calls and cards and heartfelt letters from everyone from childhood friends I hadn't seen since fourth grade to the lady who used to baby-sit for me while my parents sat in church. Even good friends I'd visited out-of-state before I was sick called just to see how I was doing.

A childhood friend from down the street—an older girl who liked it that I was game for anything, tried to teach me to play soccer, and shared my love of Eddie Rabbit's "I Love a Rainy Night"—wrote a brief but sincere letter. She told me how brave she thought I was and how fondly she remembered our adventures. Her good-bye was almost a little too schmaltzy, as if she wanted to be sure to get it in under the wire, but it was a very nice letter. I never wrote back.

The person I've known the longest amount of time in the world—our mothers were friends and pregnant with us at the same time, and we used to play in the same nursery together—wrote a very clever, yet characteristically manic, letter talking about his memories of our teenage escapades. We both enjoyed a good piece of fiction, both reading and writing it a lot during adolescence, and we let our active imaginations carry us on more than one videotaped or tape-recorded escapade of fantasy. I think I still have a copy of the soap opera script we passed back and forth for weeks, each of us adding one scene at a time; and somewhere there exists a video of the first three scenes as acted by a reluctant group of friends—one of whom was Sean. I laughed at his wildly extravagant recollection of those exploits and sat back to reminisce and laugh at us. But I never called him.

Even Sean called me every day from New York, and we'd talk about superficial things, like how I was feeling that day and his latest audition. But I rarely called him. The less I talked to him, the less I missed him. The less I said, the less I revealed about how much I desperately wanted him with me. Considering my unpredictable state, it was best to not put myself in such positions, and every call made me nervous.

Unfortunately, I almost saw all of the attention from everyone else as a bother. I felt like a movie star who just wanted to "practice her craft" and didn't understand what all the fuss was about. In some cases it bothered me that it seemed these people were contacting me simply out of a selfish need to feel as if they did something, or to be sure I was still okay even if they weren't around. Sometimes I was just too tired to think of anyone but myself. Sometimes I wanted to think of anything but myself, so I'd worry about Fiona or Joanie.

Most days I'd read a couple of get-well cards and then field two or three phone calls from well-meaning acquaintances. Some of those days, my initial reaction was to become unusually agitated. I'd put on my martyr mask, get right back on the phone, and call Fiona to complain about "people who couldn't possibly understand what I'm going through and insist on calling or writing me all the time." Fiona would sigh, probably taking a moment to decide that it wasn't the best time to fling one of her nuggets of bluntness my way. She would just say I was probably right that they couldn't understand and invite me over for the weekly screening of *The Real World: San Francisco*.

It was simple. It was easier to depend on these two friends, to throw everything questionable their way, than to reveal my neuroses or selfish concerns to anyone else. For those few months, I was convinced I didn't need any other friends because Joanie and Fiona would always understand—they'd always be there.

I hate being wrong.

I'll never know exactly why Joanie, Fiona, and I are no longer friends. I can't even remember the point in time when it happened.

One day I started getting Fiona's answering machine every time I called. One evening Fiona was unusually quiet at dinner, and she and Joanie shared knowing and exclusionary looks at least a dozen times. The next week, Joanie and I had a phone conversation that was mostly me telling her I wasn't quite sure what was going on. She just spoke cryptically and said we all had our own crises to deal with, and then asked if I'd mind if Fiona alone stood up with her at her wedding. Then I was at Joanie's wedding—sitting next to Fiona at the reception but never being spoken to, being walked away from or avoided on several occasions, listening to whispers behind my back that only clarified they were angry at me about something. After that, I never spoke to Fiona again, but I received an angry letter telling me how selfish I was and how she couldn't understand me anymore. I tried to call her back to discuss things, help her to understand and try to understand her—at least rebut her misperceptions. I've never gotten that opportunity. She never picked up the phone and never returned my calls. I spoke with Joanie only once, when I called her because I hadn't heard from either of them in months—she didn't want to talk about Fiona and barely spoke about herself. I've never spoken to them since.

I rarely talk about all of this, but when I do, the listener always asks, "Isn't it possible that you just grew apart?" Of course that's a likely option. And honestly, it surely was part of it. They had so many issues in their own lives that, once they were fairly confident my cancer was gone, they decided to focus more on themselves. But that seems almost too simple. When any friendship begins to break down, "growing apart" is usually the simplest way to describe what really happened.

The part of the situation that haunted me wasn't necessarily the fact that they went away. I won't try to save face by saying I didn't miss them at all. Don't forget I'd neglected all my other friendships in favor of them, so at first I missed Joanie and Fiona's companionship greatly. But I'll never know if "growing apart" truly was the reason for the changes, whether it was as clean-cut as a natural progression or a misunderstanding, or if my worst fears came true. Did they simply run from their friend because I'd revealed too much about what it was really like to be sick? Had the cancer turned me into someone unworthy of their friendship?

I shuddered to entertain the notion that I'd offered them the most basic humanity I had to offer anyone and it was too ugly for them to embrace. As a result, I came to believe these parts were too ugly for anyone. A part of me still does.

I still regret not answering all the concerned letters and cards I received when I was sick. Somewhere in there was an opportunity to provide someone a glimpse of what I was really going through and have it accepted and embraced. I'll never get that opportunity back.

goin' to graceland

Any survivor will tell you: If you wait long enough, people will start to forget that you had cancer. Well, they will at least begin not to think about it all the time. The good thing is that it usually happens right around the same time the survivor starts to forget too.

For the first year, I was visiting the doctor and having a CT scan every two or three months—more frequently if I felt what I thought was a symptom and spent a day worrying before making an appointment and then waiting by the phone for the results of the blood test or x-ray. After a while, the "symptoms" just became little annoyances and then part of what my body had become. I'd go for days at a time without thinking about the cancer, the risks, the fact that I should be dead and might have to fight it again very soon. I welcomed those days when no one, including me, brought up anything having to do with that—the days I didn't feel that pit of fear in my stomach.

I was beyond craving "normal," but I acquired more simple cravings, like a life without the constant presence or thought of cancer. Seems like a simple need, right? Turns out, I needed a little help, so I started to gravitate toward things that had nothing to do with being sick. I began refusing invitations from long-time colleagues, Sean and I started talking much less frequently, and, finally, I said good-bye to my support group. All of the original members had moved on or passed away, but I'd still gotten to know them pretty well—the result of sitting and listening intently and saying nothing every week. They were good, interesting people who were sympathetic in ways that non-cancer-patients or non-survivors can't be. But they were in different places—ones I needed to leave behind me. Out of a need to be away from that world, I knew I wouldn't keep in touch with them after I left, but I still thought about them often.

I'd begun making friends with people who didn't know anything about the cancer and felt that was a better support group for me just then. This was made easier by the fact that Fiona and Joanie were officially out of my life, but it also meant that the hang-ups caused by the resulting insecurities lingered. This made it all that much easier for me to avoid the "c-word" even when it might have come up. Even if my new friends did know about my health history, they hadn't known me at all while I was sick. They hadn't seen the neuroses and mania firsthand, and therefore were not free to judge me for what I was still convinced, despite therapy, were moral ambiguities and deep character faults. They just thought I was a normal person who could drink as much as she wanted and didn't have to worry about the germs from the handrail.

Of course, I very quickly latched onto these friends who didn't know me as well, but thought I was great nonetheless. One of them was a colleague, Darcy, whom I ended up working with pretty closely on a fairly large account. By then I'd moved back to

Cincinnati with Lucifer and a present from Lin and Tami, a very sweet black-and-white cat named Bootsie, and I had taken on a lot of the work for an account I'd actually fought for a few months into my reemployment, when I thought I was ready to dig back into work. Then, my body protested, and I started getting sick from the extra hours and effort. I once again had to go to my boss, the founder and head of the company, and tell him I couldn't keep up, even though months ago I promised him I was ready for a challenge. He agreed almost too quickly for my comfort, but I was glad he allowed us both to adopt a yet-again low set of expectations. I walked out of the office at once relieved and humiliated. Darcy was hired within a week.

It made sense for me to show him the ropes, so I spent at least a week working with him exclusively. Our supervisor was constantly preoccupied with her more prestigious account, so we quickly bonded over all of the grunt work we were forced to undertake.

Darcy was new to the city and adventurous yet measured, like me. He liked to talk endlessly about nearly nothing and almost everything, and was fond of asking questions and then answering them himself—"What do I think about that? I'm not sure." I appreciated the fact that he liked to turn a meal into an event, whether cooking or eating out. So, a weekly, sometimes twice-weekly, ritual was nothing more than a shared meal, accompanied by much lively, even intimate, conversations about friendships, ambitions, professional conflicts, family turmoil, the media. On a few occasions it became necessary for me to bring up my health history in order to tell a story or completely explain my opinion or experience, but I avoided that whenever possible. He never asked me to talk about it at length, and he only rarely asked me questions about it at all.

On occasion, we talked about our place of employment in personal ways, more than just who was annoying us that week.

I told him about the good and bad things that folks in the office had done since I'd been diagnosed—a benefit they'd held to raise money to make up for the company's lacking health insurance coverage and complete lack of disability benefits, raises that had passed me by even when I was working my tail off and undergoing treatments, awkward moments with the head of the agency when he revealed that, with my cancer history, I wasn't eligible for the disability insurance he finally was going to offer employees less than a year after my transplant. Eventually, he knew that these things, plus the fact that my body just wasn't up to working hard at a job I had no passion for, was why I had not been able to keep up with our now-shared account. He said all of these things didn't color his opinion of the office much, but I've always believed they did.

Darcy was good about not throwing in my face the fact that he'd picked up the slack for me when I was floundering, but that fact was always there. Every day that I walked into the office to the responsibilityless, ambitionless position I'd created, I was reminded of what I failed to accomplish. How the new guy had to save the account for us. He was my friend, my dinner buddy, but for a long time his presence alone forced me to think more about what I couldn't do than what I could.

Then I had a doctor's appointment. It was many months into my cancer-free life, and I was due for a check-up. The day of the appointment I got a little twinge in my stomach, but I didn't dread the day as I had many times before. I was beginning to feel more confident that the evil, feared menace wasn't going to reappear and take my not-so-all-about-cancer life away from me again. Dr. Reinhardt didn't need a CT scan with every appointment anymore, so he shared that confidence. I answered the questions, talked to the nurse about how I hadn't had any symptoms, talked to the doctor about how normal and truly cancer free my life was becoming, and then all that was left was the blood test. The nurse stuck the needle in, and I felt that flush that comes over my face and hands when I feel the fluid leaving my body—like life itself is being drained from me. As I looked at the blood flushing into the little chemical-smelling tubes, the blood so dark it almost seemed black, I panicked.

What was in that darkness, that fluid that only seems red when it's outside the body because the light hits it and we can see it for what it really is? Something could hide in that blood, something so tiny and transparent that it can flow through the body hundreds of times before deciding where it's going to stop and make a little tumor-shaped home. I sat, flushed and cold, and imagined those little stealthy creatures lurking about inside me, hiding from the doctors and waiting to attack, while the nurse sent the tubes to the lab for a reading. Not long after, she returned and said the lab was backed up and she'd call me with the results no later than tomorrow—it was 4:35, and experience told me I wouldn't hear a thing for at least twenty-four hours. Belligerent and manic, I went home.

So there I was, sitting at home feeling very old and filled with worries. The stack of medical bills was getting larger with every mail delivery, time was passing me by, and I really, really didn't want to go to work the next day. And there I was, waiting for a phone call that could redirect the path of my life yet again. I felt obligated to consider everything but myself and powerless to see my future yet.

Just then I got a phone call from Darcy, and I racked my brain for topics that didn't have anything to do with my currently over-whelming fear of a recurrence. We sat and talked about what we'd had for dinner, then began complaining good-naturedly about some things that'd happened at work that day. Then our true disillusionment kicked in, and we started full-on complaining. Suddenly, we weren't complaining about our employer anymore, we were complaining about all employers—calling them "The

Of course there are more rare species within the group, some of whom do not have the aforementioned secretive traits: The Walking Wounded, The Outspoken, The Outers, The Pity Seekers, and The Back Pockets, to name just a few. If you practice and become good at spotting Cancer Survivors, you will come to know these rare species and find a few offshoots of your own.

THINGS TO CONSIDER FIRST

Having had cancer is a big part of who a Cancer Survivor is, whether she recognizes it or not. While it is not something that many feel the need to carry around like a burden or a badge, or even think about that much, Survivors are most likely touched by this past in a practical or emotional way each and every day.

For most, there are continuing and continuous health concerns and risks that make it top-of-mind, especially first thing in the morning, at meal times, and as they complete their evening ablutions. For many Survivors, there are lifestyle changes that were either chosen or imposed as a result of physical changes, traumas, or realizations—exercise regimens to strengthen weakened hearts and bones, food allergies they've never experienced before, new career goals not thought of since college, many new or many fewer friends. Most of all, the majority of them still nervously return to an oncologist every three months to a year, have some tests done, and then wait around for a phone call telling them whether they're in for another fight or not. Most never really get over the anxiety of that regular visit.

All Survivors, whether six months or ten years out, have been through an emotional roller coaster of varying intensities and have changed, even if imperceptibly. These can be happy or unhappy personal changes, but they are unavoidable changes, nonetheless. The possible alterations vary as widely as the human experience, so I will only list some of the most common: more prominent focus on quality of life, diminished fear of death or sickness,

increased disdain for health insurance companies, stronger relationships with family and friends, better appreciation for really
good nurses, adjustment of career path, fear of symptomlike aches
and pains, abnormal fear of unexpected phone calls, and on and
on. These and other changes are experienced in various combi-
nations and frequently change throughout a Survivor's life.

THE WALKING WOUNDED

There are those who either deny or cannot move past their
cancerous past. These are Survivors who often are wounded
deeply by their experiences—whether a one-time operation or
many years of harmful treatments and frustrating recurrences.
Members of this group aren't as rare as you may think, but are
still difficult to observe for Survivors and non-Survivors alike.
Often they are ignored by the group because they remind others
of how they could be or, more commonly, how they once were.
And for non-Survivors, they validate the worst fears about what
it's like to go through cancer. While it is human nature to reach
out to fellow humans who are suffering and can be helped, this
doesn't always occur.

Gert, the ancient woman from my support group with a nasty
disposition, an interminable one-person pity party, a partial larynx,
and a continuing smoking habit is a perfect example of a Walking
Wounded. She was the only member of the support group,
besides J.T., I never said good-bye to.

THE PITY SEEKERS

Related to The Walking Wounded are the more deliberate
Pity Seekers. Again, Gert fits into this category, as does Joey the
colostomy-bag wearer who refused any sympathy but attended
every week so he could tell everyone how horrible he felt.

Pity Seekers are happy to tell people what they have gone
through, are going through, may go through some day. They want

their cancer survivorship to define them, not in a good way, and want to be awarded, rewarded, sympathized with, and paid attention to constantly because of it. Cancer Survivors want you to believe these are a rare species among Survivors when the truth is that nearly all Survivors have been Pity Seekers for at least a short period of time. This does not make them bad people, though they may believe otherwise.

THE OUTSPOKEN

Some Survivors have entire speeches—entire books—full of cancer stories to tell. These are a bold, aggressive, risk-taking species, but not always the smartest. Truthfully, the motivation for writing or speaking about their experiences is often altruistic, but not always, by far. It is said that a diary is cheap counseling. Well, taking it one step further is the ultimate therapy for some. Validation, support, acceptance, risk, and the closet thrill some get from sharing their most intimate secrets are all benefits that aren't obvious or even shared by many outside The Outspokens.

There are increased dangers these Survivors face. Frequently, they are pegged as either self-indulgent or immeasurably self-sacrificing. While a portion of each of these traits can be attributed to this group, as well as to many Survivors outside of the group, neither extreme can be fully attributed.

There are other dangers for all Survivors that are increased for those from The Outspoken group. It is never a good idea for Survivors to share too readily or eagerly that they have cancer in their files. Even those who aren't afraid of sharing typically don't walk around telling stories willy-nilly. While on a personal basis most people find the stories of Cancer Survivors to be somewhere between interesting and inspiring, an employer, a business partner, a health insurance provider, and a potential romantic partner are among those who may not find the trait desirable without more in-depth knowledge.

When I try to put myself in a non-Survivor's shoes, I imagine it must be sort of like finding out that the beautiful house you just bought has major plumbing problems and could suddenly go back on the market without warning. Survivors understand that not everyone will view it as impersonally as that, but in a practical, bottom-line-oriented world, it absolutely is not outside the realm of possible reactions. Therefore, it is almost always the Survivors' survival instinct to withhold this sort of information until they feel there is minimal risk in revealing it.

THE OUTERS

Related to The Outspoken are The Outers. Some feel it's in all Survivors' best interests to talk openly about their experiences—demystification and destigmatization of the disease and all that. Some go so far as to "out" others without knowing the Survivor's comfort level or intentions. They justify this somewhat extreme reaction because they want their fellow Survivors to experience a bond they've felt that perhaps the out-ee is too shy to take advantage of.

Ginnie told me a story once about sitting on an airplane with a woman and, in her own bold Ginnie way, striking up a casual conversation. The woman was wearing a white cotton button-down blouse that was opened to a couple of buttons down, and Ginnie caught a glimpse of her front shoulder blade and a few inches around it. Immediately, she recognized it as the small round protruding disk of a port-o-catheter and didn't even hesitate before saying loudly with a smile, "You have cancer!" as if she were discovering she was her long-lost best friend from kindergarten. Ginnie said the woman's eyes got kind of big, and she glanced around to see who'd heard the declaration before affirming in a low, inside voice, "I did. It's been three months." Ginnie began talking excitedly about what kind she had and how she knew there was a special bond between them when they started talking and suddenly asked how she was doing. The previously open woman sat and listened to Ginnie's stream-of-consciousness purge, but only smiled timidly and provided minimal details about her treatments, subtly urging Ginnie to keep her voice down. Their two-sided conversation was over, and Ginnie didn't even notice or feel badly until the plane was landing. She never apologized to the woman and, to me a few days later, justified herself on the grounds that the airplane companion "should want to talk about it after all." I told her she should have apologized.

You see, Ginnie, like other Outers, only saw good in others knowing. While I agree that we all benefit from Survivors' stories, I also believe that only the "others" benefit from stories the Survivor wishes to remain private about.

THE BACK POCKETS

The vast majority of Cancer Survivors fall into this category. Their experiences, like most of their life's experiences, are with them, but tucked away neatly where they are more easily dealt with, like a back pocket.

Personal example: Some people asked me, not long after I was pronounced cancer-free, if I thought about my cancer all the time . . . if it was a part of everything I did now. I could honestly answer "no," though I didn't qualify it as I wanted to. No, I don't sit around and pine about my experiences. I haven't devoted my life to the fight against cancer or helping patients going through therapy. In fact, days, even weeks, go by without anything but a few fleeting thoughts or memories related to it.

However, I didn't mention that I thought about it when I was taking a shower, as I washed skin buckled by biopsy and catheter entry-wound scars. Or, after my shower, when I took my estrogen pill to replace the hormones lost by chemotherapy-induced menopause. Or, as I pulled out the bottle of sanitizer in my purse after shaking hands with a colleague with a cold because my immune system is naturally on the low side, and I most likely will catch the cold anyway. Or, as I watch the evening news capped off by a story about another long stride the cancer therapy or research community has made.

Cancer Survivors, like everyone, have memories from a portion of their lives that can never be erased. Just like the memory of your high school prom, whether you stayed home out of protest or were the prom queen, there always will be a memory associated with it. It will never be erased because it was important in your life at the time. Even though you may rarely think about the experience, the memory of it may still cause great joy or great pain even many years later—and there may be several things that occur throughout each year, month, or day to remind you of it.

Depending on what generation you were born into, you probably remember what you were doing when Pearl Harbor was bombed, on VE-Day, when JFK was shot, when the space shuttle Explorer exploded with Christa McAuliffe onboard, when JFK Jr. crashed his airplane off the East Coast, when three airplanes hit the World Trade Center and the Pentagon within an hour of each other. People all over the country were fighting for their lives or having babies or getting fired at the same time as these events. And millions of people call September 11th their birthday. I will never forget I was in a car accident in a parking lot the same day.

What were you doing during the O.J. trial? I was living with my parents, preparing for a bone marrow transplant and, consequently, spending a lot of time avoiding the television. Where were you when the Oklahoma City bombing occurred? I was standing and staring at the television in the lobby of the outpatient oncology ward in the hospital where only a week before I'd been a patient in the bone marrow transplant unit, waiting for a blood transfusion, my mouth hanging wide open with shock.

The world doesn't stop just because someone is sick. And those memories, or their context, can't just be erased.

These are just things to keep in mind as you consider your approach.

THE APPROACH: TIMING IS EVERYTHING

With all of these variables to the "reveal" aside, there are still things to consider once cancer is occurring or has occurred. Even if one's cancer history does come up in a completely natural way, and both people are deemed trusting and trustworthy enough to handle it, there are a multitude of things that could happen next. The most common reaction from the listener is surprise—something between "oh, gosh" to dumbfounded shock, but usually on the lower end. Because of this, there is at first at least a mild amount of discomfort for one or both parties, especially if the conversation didn't seem to be veering in that direction in the first place. Cancer, along with the Nazi party and quantum physics, is one of the ultimate conversation stoppers, and another reason why Survivors often put off the "reveal" a while.

It is important to note that while Survivors may not offer their health history to a new acquaintance, or even a recently made friend, most are proud of their status and have, at minimum, a few nuggets of something to offer about what they learned as part of the experience. Most will only offer more information if asked, and most don't mind being asked once the truth is out there. This poses a couple of quandaries, however. First, because most non-Survivors will not ask many questions without prompting, and most Survivors will not offer information unless asked, the usual result is that little information, usually very superficial, is passed between Survivor and inquirer. A communications predicament, if you will. Second, it isn't necessarily true that all Survivors have actually gained their nuggets of wisdom through their own experiences, even if they claim them as their own. Some have had to come up with something even if not based on their own experience, simply because people will always ask and are surprised if there is no satisfactory response. For most Survivors, this tendency to inquire is not received with hostility. It is merely an environment to which Survivors have learned to adapt. It is one of the mysteries of the species.

Truly, timing is everything. If it's too early to tell a new friend or colleague, you come across like that old lady at the bus stop who turns to you and tells you she had a bunion removed that morning, or the drunk at the bar who reveals that he has only one testicle. It's just too much casual sharing to be comfortable. But there can be countless times that something doesn't make sense until the Survivor reveals this most personal fact. Often, a question is answered more clearly, or a story or memory is more easily told, if the teller injects this fact. Sometimes it seems almost like lying, or at least deception by omission, if it's not included.

Timing aside, there are so many other things to consider. Once the cat is out of the bag, it's out forever. The listener probably will never look at the Survivor in the same way again. Sadly, there's a realistic fear that the image or impression the Survivor has worked so hard to gain could disappear—possibly swaying it unfairly negatively or unwarrantedly positively—in one declarative sentence. Hence, the ongoing dilemma for any kind of Survivor: How and when do I tell people about it? Do I tell at all? Will they go away? Should I worry about their reaction at all? These are important questions to Survivors and, for me, the answers keep changing.

Personal example: When I was declared cancer-free—like sugar-free except with more flavor—other Survivors and people who were well aware of my health history surrounded me. There was no need for eluding the facts. In fact, they usually brought the

subject up before I did. The fact that I'd survived was more mirac-
ulous to everyone else who didn't watch the process step by step
and get used to the idea one tiny increment at a time.

But inevitably, situations arose—where I knew someone
didn't know about it and something wouldn't make any sense
unless I explained. I avoided it as much as possible. The conversation inevitably veered off its path, and the spotlight became permanently positioned on my face while I spouted the hundred-
words-or-less explanation of the last year and a half of my life.
Normally, the conversation never returned to the interesting place
it was going before the "c-word" came up. Just think of the dozens
of conversations and keen insights I could have benefited from—
all lost now to that pesky conversation killer.

I am positive that a woman I shared an apartment with a
few years later was one or two sentences away from finally
telling me she was gay when the subject of a doctor's appointment I'd had that day came up, and she began asking questions.
She didn't come out to me until two months later and only after
she'd broken the ice by asking more questions about the cancer
for an hour and a half.

The roommates I'd had before that, a musician couple that
"managed" an apartment with a long row of bedrooms filled with
people paying part of their rent and sharing their kitchen, still do
not know. There simply was no reason to tell them, and, between
their recitals and screaming matches, it never came up.

MATING

Since most Survivors are older, married, or committed and
have already had children, the issue of mating is not necessarily
a critical one for the group as a whole. For those who are young
and uncommitted, it was, and is, a very critical one. The first issue
to arise regarding the mating rituals of Survivors is, again, the
"reveal." When, how, and to whom can make all the difference